FUTURE BIOETHICS

OVERCOMING TABOOS, MYTHS, AND DOGMAS

FUTURE BIOETHICS

OVERCOMING TABOOS, MYTHS, AND DOGMAS

RONALD A. LINDSAY

PENTAGON PRESS

Future Bioethics: Overcoming Taboos, Myths, and Dogmas
by Ronald A. Lindsay

ISBN 978-81-8274-365-6

First Indian Edition: 2009

Original English edition published by Prometheus Books,
59 John Glenn Drive, Amherst, New York 14228-2119.

First published in India by
PENTAGON PRESS
206, Peacock Lane, Shahpur Jat,
New Delhi-110049
Phones: 011-64706243, 26491568
Telefax: 011-26490600
email: rajan@pentagon-press.com
website: www.pentagon-press.com

Printed at New Elegant Printers, New Delhi.

For Debra, my west wind

CONTENTS

Preface

Despite the strong interest that many appear to have in controversies such as legalization of assistance in dying, funding of stem cell research, conscientious refusal by pharmacists to dispense emergency contraception, the regulation of genetically engineered plants, and similar topics—which are usually grouped under the heading of "bioethics"—there are few books that engage the reader in a search for the appropriate public policy on these issues. Books on bioethics tend to fall into two categories: those written primarily for scholars working in the field and those written for the general public. Books in the latter category may make for livelier reading, but many of them share a common deficiency. They are long on assertions and short on reasoning. They rely heavily on rhetoric and slogans, such as the "sanctity of life" or the "right-to-die." As a result, they ultimately fail to present a convincing case for whatever policy or policies the author may be advocating. In addition, and perhaps more important, they fail to involve the reader in a thoughtful consideration of policy options.

Books written for fellow scholars are usually more tightly reasoned but fail to attract attention from the general public because they rely on background knowledge that may not be immediately accessible to the person who is not a specialist. They may influence fellow scholars, but they do little to help the interested layperson in coming to her own decision about appropriate public policy. This is regrettable because these issues affect all of us, and in a democracy, where the ballot box may be the ultimate arbiter of public policy, an informed citizenry is critical.

My aim in this book is to present arguments that are accessible to any person who is interested in the controversies discussed in this book. For each of the policy issues that I address, I make clear at the outset the factual and moral premises that I will utilize in my reasoning and I then invite the reader to reason along with me. Of course, I hope to persuade the reader that the policies I advocate are the ones that we should adopt, but even if I fail in this objective, it will at least be clear to the reader where we disagree with each other. That in itself will be a welcome outcome given that with respect to these controversies, people seem to often talk past each other, with mutual incomprehension rather than understanding being the result.

I believe one reason the issues discussed in this book attract such controversy, and also seem to resist resolution, is because they are the product of a revolutionary transformation in biomedical technology in the past few decades. We have to confront issues that prior generations did not have to face—and may not even have imagined. Not unexpectedly, these novel issues lead to moral confusion. Some try to resolve these issues by clinging reflexively to moral beliefs that were developed under different social conditions and remain anchored in the past. This is the wrong approach to take. Unthinking adherence to such beliefs transforms them into taboos, myths, and dogmas that impede our progress in developing and implementing sound public policy. In this book, I will show how to overcome these misguided beliefs and patterns of thought.

The issues discussed in this book are arranged in rough order of their appearance in the arena of public policy disputes, beginning with assistance in dying—an issue that has been debated now for a couple of decades—and proceeding to issues relating to enhancement technologies and stem cell research. This arrangement is no accident. In assessing my arguments, I believe it will assist the reader to start with issues with which he may have more familiarity before proceeding to more recent controversies.

The title of this book has a double meaning. It not only indicates that resolution of the issues addressed in this book will shape our future, but it also indicates that I am advocating an approach to resolving controversies in bioethics that can be applied to other controversies that are likely to arise as the biotech revolution continues. For this reason, although I recognize it is tempting to skip any chapter that discusses methodology, I urge the reader not to pass over chapter 2. It provides a framework for resolving public policy disputes in bioethics. Plus, it may be the only chapter in a bioethics book that discusses the film version of *The Last of the Mohicans*.

Although this book is directed at the general public, the arguments set forth (many of which first appeared in some of my articles in scholarly journals) are intended to be rigorous enough to withstand scrutiny by specialists in bioethics. Writing a book that is accessible to the general public that will also pass muster with fellow scholars is a task easier stated than accomplished. One runs the double risk of being too obscure for the general reader while simultaneously being too imprecise for academic critics. Whether this book succeeds in its objective or succumbs to these risks is a judgment you will have to make.

Acknowledgments

This book would never have seen the light of day without the encouragement of Paul Kurtz, professor emeritus of philosophy at the State University of New York at Buffalo. It is not only that Paul recommended I submit a manuscript to Prometheus Books; throughout our acquaintance of more than twenty years, he has unfailingly urged me to pursue my love of learning, and to write whenever possible. If Paul holds anything sacred, it is scholarship.

In part because of Paul's encouragement, I pursued a PhD in philosophy (with a concentration in bioethics) during the early 1990s, at a time when I was already working as a lawyer. I owe much to the education I received at Georgetown University, and my intellectual debt is especially large to Tom L. Beauchamp, professor of philosophy at Georgetown and Senior Research Scholar at the Kennedy Institute of Ethics. Tom was the mentor for my dissertation, which was on assisted suicide and euthanasia. Tom cannot abide shoddy reasoning and sloppy writing, and I hope this book has benefited from the training I received from him.

Ruth Mitchell did a masterful job in assembling the bibliography and in suggesting wording changes at various points throughout the manuscript.

Persons who, at one time or another, have read either portions of my manuscript or previously published articles or papers on which some of the manuscript was based include Carol Mason Spicer, Rebecca Dick, David Koepsell, Marc Spindelman, Danny Horowitz, Steve Lowe, Debra Robinson-Lindsay, and Toni Van Pelt.

As indicated, a few passages from some of my previously published works have been incorporated, in modified form, in this book. I thank the editors and publishers who provided permission to incorporate passages from the following articles and papers:

A Call to Legalize Physician Assistance in Dying for the Terminally Ill: Lessons from the Oregon Experience. Washington, DC: Center for Inquiry, 2008. http://www.centerforinquiry.net/advocacy. Copyright © 2008 Center for Inquiry.

"Enhancements and Justice: Problems in Determining the Requirements of Justice in a Genetically Transformed Society," *Kennedy Institute of Ethics Journal* 15, no. 1 (2005): 3–38. Copyright © 2005 Johns Hopkins University Press.

"Hastened Death and the Regulation of the Practice of Medicine" (coauthors Tom L. Beauchamp and Rebecca Dick), *Washington University Journal of Law and Policy* 22 (2006): 1–28. Copyright © 2006 Washington University.

"The Need to Specify the Difference 'Difference' Makes," *Journal of Law, Medicine and Ethics* 30 (2002): 34–37. Copyright © 2002 American Society of Law, Medicine and Ethics.

"Should We Impose Quotas? Evaluating the 'Disparate Impact' Argument against Legalization of Assisted Suicide," *Journal of Law, Medicine and Ethics* 30 (2002): 6–16. Copyright © 2002 American Society of Law, Medicine and Ethics.

"Slaves, Embryos, and Nonhuman Animals: Moral Status and the Limitations of Common Morality Theory," *Kennedy Institute of Ethics Journal* 15, no. 4 (2005): 323–46. Copyright © 2005 Johns Hopkins University Press.

Stem Cell Research: An Approach to Bioethics Based on Scientific Naturalism. Washington, DC: Center for Inquiry, 2006. http://www.centerforinquiry.net/StemCell.pdf. Copyright © 2006 Center for Inquiry.

"When to Grant Conscientious Objector Status," *American Journal of Bioethics* 7, no. 6 (2007): 25–26. Copyright © 2007 Taylor & Francis Group, LLC.

"Why Should We Be Concerned about Disparate Impact?" *American Journal of Bioethics* 6, no. 5 (2006): 23–24. Copyright © 2006 Taylor & Francis Group, LLC.

I would also like to acknowledge the able assistance of the staff of Prometheus Books, including, but not limited to, Steven L. Mitchell, editor in chief; Peggy S. Deemer, senior editor; Christine Kramer, production manager; and Melanie Gold, copy editor.

Finally, I would like to thank the members of my family for their patience and understanding as I devoted nights and weekends to this book. Debra, my wife, and David and Anne, my children, also provided unsparingly critical comments on various portions of the book, which improved the final product.

Debra deserves special thanks. Those who have seen me at a keyboard—if they do not immediately avert their eyes in horror—may wonder how I can get anything produced in a timely manner. Fortunately, Debra has been gracious in providing me with typing assistance for large segments of the manuscript.

1.

The Problem and the Cure

This is a book about issues that directly affect your life and the lives of your loved ones. For example, among the issues discussed in this book are whether we should legalize physician assistance in dying for terminally ill patients, whether we should allow pharmacists to refuse to dispense medication to which they have an objection, and whether the federal government should fund embryonic stem cell research—research that may produce valuable therapies but will require use of human embryos.

Controversies such as these have received an enormous amount of attention in recent decades. It is not difficult to understand why. First, as indicated, how these disputes are resolved will have significant consequences for all of us. Second, to a large extent, these controversies present ethical questions that prior generations did not have to confront. In fact, bioethics as a specialized discipline has been around only since the early 1970s. (I use the term "bioethics" as it is commonly understood, that is, a field of study focusing on ethical and legal controversies arising out of healthcare, the practice of medicine, and biomedical research.) One of the principal controversies that helped spur the development of bioethics was the litigation over the withdrawal of treatment from Karen Ann Quinlan, the unfortunate New Jersey woman who fell into a persistent vegetative state in 1975.[1] The condition of a "persistent vegetative state" had itself been expressly recognized and defined only a few years before. Prior to that time, there was no need to define that condition or to resolve controversies over maintaining someone in that condition because the technology to keep such patients alive indefinitely simply was not available.

Similarly, medical procedures and treatments such as in vitro fertilization, organ transplantation, "morning-after" contraception, and gene therapies are all relatively recent developments and have all sparked ethical and legal disputes.

The novelty of some of the ethical and legal issues presented by the recent rapid advances in biomedical technologies is underscored by the fact that both federal and state governments have created commissions and other advisory bodies to study and discuss these issues and recommend policies. For example, the Quinlan case and other controversies over patient care resulted in the creation in 1978 of the President's Commission for the Study of Ethical Problems in Medicine and Biomedical Behavioral Research.[2] Prior to the 1970s there was scant precedent for the creation of government advisories to study *moral* as opposed to social or economic issues.*

Nonetheless, despite the development of bioethics as a specialized discipline and the investment of substantial public resources to study bioethical issues, all too often our policies continue to reflect myths, taboos, and dogmas rather than reasoned analysis and an appropriate understanding of the implications of these new technologies. Part of the explanation for this is again attributable to the novelty of these technologies. There is an aversion that many instinctively feel to the new and unfamiliar. Virtually every new medical procedure and technology has been opposed on the ground that it is somehow "unnatural." Those old enough to remember the 1960s may recall the shock and horror experienced by some at news of the first heart transplant; many condemned the procedure as unnatural and inherently immoral.[3] Today, few would oppose heart transplants, or organ transplants in general, as immoral. (The current ethical debates surrounding transplants usually focus on legitimate ways to encourage organ donation.)

Although one frequently sees the terms "natural" and "unnatural" thrown around in debates about bioethical issues, these terms actually explain little and obscure much. They function as substitutes for mean-

---

* I will treat the terms "ethical" and "moral" as synonymous. In some contexts distinctions are made between these terms, but for my purposes there is no need to draw a distinction.

ingful reasons and arguments. Almost anything can be described as "natural" or "unnatural." Using wheeled vehicles as a mode of transportation and obtaining milk from cows are unnatural actions in the sense that our human ancestors were not born with rotating disks on their feet nor were they initially in a world in which tame animals supplied some of their nourishment—and the use of these techniques were undoubtedly opposed at first by some who found these innovations frightening or abhorrent. (Debating point from 10,000 BCE: "If we drink milk from cows, we're on a slippery slope to making cheese.") The point is that saying something is "unnatural" is simply a superficially more impressive way of saying, "I don't like this, but I can't tell you why." Yet new technologies continue to be opposed by some merely on the ground that they are somehow unnatural.[4]

A related objection to new biomedical procedures or techniques is that they require us to "play God." How often have we heard that phrase in the last forty years? Any attempt by humans to control and shape their lives in ways not previously contemplated by some religious tradition results in the claim that we are trying to "play God." This objection appeared when birth control pills were introduced, when organ transplants were first attempted, when patients were first placed on or removed from respirators, and so on. But this objection is not very informative; indeed, it is not clear whether it has any specific meaning at all or is simply an emphatic way of indicating opposition. To the extent this objection has any core meaning, it seems to suggest that it is improper for us to use our own cognitive and emotional resources to come to a decision about right and wrong. Certain decisions are God's prerogative. If that is the meaning of this objection, then it is incoherent, because every time we do something because it is morally required or permissible and every time we refrain from doing something because it is morally impermissible, we "play God." This includes both the acceptance and rejection of new biomedical procedures or techniques.

We cannot avoid making decisions on an issue allegedly reserved for God because a decision not to undertake an action is itself a decision. If I decide not to undergo a heart transplant because undergoing such a procedure would be to "play God," I am playing God by deciding to forego the procedure. Even if we assume there is a God and that God takes an interest in human affairs, we have no option but to rely on our own reason, understanding, and experience to determine what is and is not morally acceptable. This is not usurping God's authority. If there is a God, we are simply using the means that She/He/It has given us to consider and resolve moral problems.

Some readers may react negatively to this assertion and claim that we know something is good because God commands it and something is bad because God forbids it. But how do we know when God has commanded something or forbidden it? If someone informed you that God has commanded us to kill every other female newborn and shoved some "holy" writings in your face to prove this, would you accept this supposed commandment? Obviously not. Presumably you would consider this person to be morally deranged. If God is supremely benevolent and would only require us to do what is good, as nearly all religious persons claim, then as a matter of logic, we must first determine what is morally good and bad *before* we decide what constitutes an authentic commandment from God and what represents a counterfeit revelation. There is no shirking your moral responsibility. In rejecting (or accepting) some new technology or procedure, you may pretend that you are only following God's commandments, but you have at least implicitly made a decision about which moral principles are good and, therefore, endorsed by God's commandments. Either that or you are simply slavishly following someone else's interpretation of what constitutes a commandment from God.

Nonetheless, the conviction held by many that morality must be based upon religion and that we can discern what is morally acceptable by consulting religious writings or dogma results in deeply misguided attempts to mine the Bible, the Koran, or other writings deemed sacred for insights on bioethical controversies. This occurs repeatedly despite the fact that these writings were generated at a time when there were no respirators, no organ transplants, no gene therapies, no embryos available for research, and so on. Not surprisingly, these writings are utterly silent on the issues presented by these innovations, unless these writings are very imaginatively interpreted—and then we are again back to the question of what constitutes a correct interpretation of God's word. Blind adherence to religious dogma has proven a serious impediment to progressive policies, as I will demonstrate in this book.

However, this book should not be interpreted as some sort of thinly veiled attack on religion. Given the popularity of some recent books with that theme, perhaps it would improve sales if it were, but that is not my focus. This is a book about bioethics, not metaphysics. Many bioethicists who are religious have made valuable contributions to moral inquiry. They have done so by offering reasons, not religious dogma, for their positions—reasons that can be assessed and evaluated by anyone regardless of their religious background. Just as someone's religious beliefs or lack thereof are a bad predictor of that person's character (there are thoroughly odious atheists just as there are thoroughly odious Christians, Muslims, Jews, etc.), similarly, religious beliefs or lack thereof are not necessarily correlated with an insightful understanding of bioethics.

Certainly, it would be incorrect to infer that instinctive rejection of innovation and misguided reliance on religious dogma are the only impediments to ethically sound bioethics policies. Secular ethicists can have ideological commitments that result in stubborn adherence to positions that are incoherent, irrational, or devoid of factual support. One example is the assertion that if a policy might have a "disparate impact" on some allegedly vulnerable group (typically, women, racial minorities, and the disabled are cited as examples of vulnerable groups), then this by itself is a sufficient reason to reject this policy. ("Disparate impact" here means that the policy has a disproportionate adverse effect on some group, such as African Americans being denied appropriate medical treatment more often than whites.) Invoking disparate impact as the basis for objecting to some policy is very fashionable among bioethicists, to the point that all too often they do not consider whether a policy's alleged disparate impact actually provides a reasonable basis for objecting to this policy. The classic example is the claim that we cannot legalize assistance in dying because it would have a disparate impact on women and racial minorities since members of these groups are more likely to be pressured into requesting assistance in dying. Leaving aside the point that there is zero empirical evidence to support such a claim, does such a claim even make ethical sense? Effectively, those who raise this objection are implying that if white men were pressured into choosing assistance in dying at rates that reflect their proportion of the population, then it would be permissible to legalize assistance in dying. But this is absurd. A coerced request for assistance in dying is morally objectionable no matter what the race or gender of the person being coerced. Evidence that we cannot legalize assistance in dying without placing large numbers of individuals at risk of coercion would be a legitimate argument against legalization, but it is the overall number of individuals who are placed at risk, not their race or gender that counts.

For the issue of assistance in dying and for all bioethical controversies, we need to have a clear understanding of the rationale for our policies. All too often we fail to ask the fundamental questions: What precisely are we trying to accomplish through our policies? How will our policies further human interests? Do our policies strike an appropriate balance between the rights of the individual and the community's interests? Are our policies consistent with our policies and ethical views on other issues?

In this book, I will analyze the flaws in the various taboos, myths, and dogmas that continue to influence our policies—to the detriment of our health and well-being.

I will begin with a topic that has been the subject of much debate in the United States and in other countries, namely, the legalization of assistance in dying.[5] Despite a decade of evidence from Oregon (where assistance in dying has been legal since 1997) that confirms legalizing assistance in dying does not place patients at risk, opposition to legalization remains adamant in some circles. This opposition has no basis in reason or fact; in the final analysis, it is predicated on a taboo mentality. Several of the principal arguments against legalization—based on the alleged inability to control abuses, the sanctity of life, and the disparate impact that assistance in dying will have on vulnerable groups—will be examined and found wanting. In the process, I will show the profound inconsistency between our policies on cessation of life-sustaining treatment, which any competent patient may refuse, and our policies on assistance in dying, which remains illegal outside of Oregon. I will devote more space to this topic than others because the flaws in reasoning of the opponents of assistance in dying are indicative of the more common flaws in bioethical arguments generally.

In recent decades, federal and state laws have provided that various healthcare professionals, ranging from physicians to nurses' aides, can refuse to participate in certain healthcare procedures based on their "conscientious objection" while still retaining their jobs. The latest group to join this parade consists of pharmacists who refuse to dispense certain forms of contraception. Are healthcare professionals there to help patients or are patients there to secure a livelihood for individuals who want to be able to pick and choose the services they provide and obstruct patients' choices? I will demonstrate why the category of "conscientious objector" is misapplied in almost every case involving healthcare professionals.

Many individuals suffer from the misperception that organic food is not only healthier for them, but is also morally preferable to food cultivated with the use of conventional methods or scientifically enhanced methods, such as genetic engineering. But organic farming methods typically consume more land than other methods, resulting in more harm to the environment and feeding fewer people. In addition, no study has demonstrated that organic produce is overall healthier for humans than conventional or genetically enhanced produce. Some of the attraction to organic foods is based on misinformation. Many consumers believe organic foods must be grown without pesticides. Wrong. Organic farmers may use pesticides derived from organisms and various minerals—you see, these pesticides are supposedly "natural." If the misperceptions that many have about organic farming resulted simply in a mistaken understanding of its methods, this probably would not be a significant issue. However, myths about organic farming also contribute to unjustified suspicion of food produced in part by bioengineering. This is a serious issue because calls for banning or onerous regulation of bioengineered products not only lack any scientific justification, but prevent our reaping the benefits of these new technologies, which could result in greater agricultural yields and a substantial decrease in the use of pesticides. Fears and fantasies about the "unnatural" pose a much more serious threat to your well-being than genetically engineered corn.

The bias against the so-called "unnatural" is also evident in the blanket opposition of many to pharmacological and genetic enhancements being developed for humans, such as modafinil, a drug that can suppress the need for sleep and improve certain cognitive skills. Many have called for a ban on, or strict regulation of, such enhancements, even though few truly significant enhancements have been developed as of yet. These proposed prohibitions are imprudent at best and threaten our development of useful therapies, which rely on the same technology and research as enhancements. Furthermore, outside of special contexts, such as athletic competitions, there is no meaningful ethical distinction between self-improvement through currently accepted ways of enhancing an individual's capacities, such as by education, conditioning, and the like, and enhancements through genes and drugs. The concern that the wealthy will have an advantage in terms of access to enhancements does not, by itself, warrant a ban on pharmacological or genetic enhancements. We do not ban private schools merely because the wealthy have greater access to their perceived benefits.

The debates in recent years over the use of embryos in stem cell research provide the clearest demonstration of how taboos and dogma can cloud our thinking. Opponents of stem cell research contend that the embryo has the status of an adult human being, but their arguments for this claim are undermined by common sense and their own inconsistencies. They claim that an embryo is a human person because: it has a soul; or it has the genetic composition of other humans; or it has the inherent potential to acquire the capacities of a human person. But the soul argument is religious dogma pure and simple. The genetic composition argument assumes not only that genes have some intrinsic value, but blithely ignores the fact that many we regard as humans (e.g., individuals with Down syndrome) do *not* have the same genetic composition as other humans (they have an extra chromosome). Finally, with respect to their potential, embryos do not develop on their own; they need a uterus. The potential for an embryo that is not implanted in a uterus to become a human being is equivalent to zero. Moreover, if embryos have the same status as adult humans, why do we allow IVF clinics to throw "spare" embryos in the trash?

Progress in bioethics has been blocked long enough. We desperately need a well-reasoned, pragmatic approach to controversies in bioethics that avoids reliance on taboos, myths, and dogmas, whether these result from religious or ideological beliefs. This book will help lay the foundation for that effort.

NOTES

1. The New Jersey Supreme Court ultimately ruled that Quinlan's ventilator could be removed. *In re Quinlan*, 70 N.J. 10 (1976). The court's decision is appropriately recognized as groundbreaking, as prior to this decision there was no generally recognized right to refuse or direct the withdrawal of life-sustaining treatment. The irony is that Ms. Quinlan lived for about another ten years after the removal of the ventilator. Apparently, the nuns who were taking care of her had weaned her from the ventilator in anticipation of the court's decision.

2. This commission ultimately published several important reports, including *Defining Death: Medical, Legal and Ethical Issues in the Determination of Death* (Washington, DC: GPO, 1981) and *Deciding to Forego Life-Sustaining Treatment* (Washington, DC: GPO, 1983). Another government commission established in 1974, the National Commission for the Protection of Human Subjects of Biomedical and Behavioral Research, is usually given credit for being the first bioethics commission. Its report focused on the regulation of medical research involving human subjects. *The Belmont Report: Ethical Principles and Guidelines for the Protection of Human Subjects of Research* (Washington, DC: GPO, 1979).

3. For an overview of some of the initial objections to heart transplants, see Kenneth Vaux, "The Heart Transplant: Ethical Dimensions," *Christian Century* 85 (1968): 353–56. See also Robert M. Veatch, *Transplantation Ethics* (Washington, DC: Georgetown University Press, 2000), pp. 1–12 (discussing some religious objections to organ transplants).

4. In fairness to those among our ancestors who may have opposed the domestication of animals, I should point out that there were biological consequences arising out of this innovation. Many of our diseases, such as measles and smallpox, probably had their origins in microbes present in animals. These microbes were transferred to us when animals such as cattle, pigs, and sheep became domesticated. See Jared Diamond, *Guns, Germs, and Steel* (New York: W. W. Norton, 1999), pp. 206–208.

5. Assistance in dying is also known as assisted suicide, although that terminology can be imprecise and misleading for reasons discussed in chapter 3.

2.

How to Decide Which Policies to Adopt

Method in bioethics is critical. We cannot hope to achieve a consensus on which policies to adopt if there is no agreement, at least in general terms, about the procedures we should use in addressing and resolving policy disputes.

The reader may think that this is an impossible task given that ethical disputes are notoriously resistant to resolution. Arguments about right and wrong are at the center of almost all our policy disputes, and the divisions between different viewpoints often seem unbridgeable. Just consider how sharp and long lasting the disputes have been regarding abortion.

Nonetheless, although I certainly have no illusions about persuading everyone, I believe we can find common ground on some issues by choosing an appropriate staring point, applying some common sense reasoning, understanding and making explicit the underlying rationale for our norms, and recognizing the proper relationship between morality and the law. This last point is particularly important because it is all too often overlooked. An action may be morally wrong but unaddressed by the law (think of everyday lying, workplace backstabbing, or rudeness). Conversely, an action may be considered morally justified, but legally prohibited (think of civil disobedience). In crafting public policy, we need to be cognizant not only of the dictates of morality but also of the manner in which the law—often a blunt instrument—can be used prudently to guide and limit our conduct.

COMMON MORALITY

But where to begin? I do not think we can begin by trying to set forth an all-encompassing moral theory that purportedly will serve as a reliable guide to right and wrong and then persuade everyone to adopt this theory. The record of success for such efforts is pretty dismal. Among professional philosophers, perhaps the two primary competing ethical theories (there are many more than two theories—probably as many theories as there are philosophers) are utilitarianism and deontology, or rule-based ethics. Utilitarianism tells us to consider how our actions can maximize utility, which is then variously defined. In other words, it instructs us to focus on the consequences of our actions. Rule-based ethics maintains that certain actions are forbidden regardless of their consequences. Utilitarians and deontologists have been pointing out the flaws in their rivals' theory for about two hundred years; there have been few converts. This is not to say that there have not been important and insightful points made by ethicists in both camps, but neither side has provided a comprehensive argument that everyone finds convincing and compelling, nor is either side likely to produce such an argument. Certainly, in the space of a few pages, I doubt if I could accomplish this task, even if I were inclined to make the attempt.

But if we do not start off with a theory, where do we start? I recommend we start with what is often referred to as "common morality." Common morality represents the moral norms that virtually everyone accepts. Our moral disagreements tend to hide the fact that there is substantial agreement in our culture, and other cultures, about core moral norms. It could not be otherwise if in fact we are to live together successfully. In our day-to-day dealings with one another, we accept certain behaviors as given. We assume most people will not steal from us, break their promises, maim or kill us, and so on. Society would collapse if we could not rely on others to conduct themselves appropriately *most* of the time. And almost all of us are in agreement about what is appropriate conduct on most occasions. No morally serious person disagrees that, all other things being equal, it is wrong to steal, break promises, injure others, and so on. This is not to say that we all abide by the core norms on all occasions. Of course not. We are sometimes tempted to violate accepted norms when it is to our advantage and we sometimes yield to this temptation. But these variations in our conduct do not imply that we do not accept these core norms.

Many philosophers have identified commonly accepted moral norms or rules; although their lists differ slightly, there is substantial overlap. For example, Tom L. Beauchamp, a noted bioethicist at the Kennedy Institute of Ethics, has suggested that these core norms include norms prohibiting killing, causing pain, stealing, and punishing the innocent, and norms requiring telling the truth, keeping promises, nurturing the young, preventing harm from occurring, and rescuing persons in danger.[1] Beauchamp maintains that the common morality also recognizes certain core virtues, such as honesty, integrity, conscientiousness, and kindness. Moral philosopher Bernard Gert provides a somewhat different, but substantially similar, list of ten moral rules, along with a number of commonly accepted moral ideals.[2]

Of course, core moral norms are both few in number and very general. They must be if they are supposed to be norms that have been and are accepted by all cultures. However, in addressing a current ethical dispute they can be supplemented by commonly accepted or scientifically supported factual beliefs as well as norms that arguably are accepted widely only in contemporary culture or some contemporary cultures. As an example of the latter type of norm, think of a norm that requires us not to interfere with a person's decisions about matters important to that person, such as her choice of a career or spouse or a decision whether to bear children. It is doubtful that such a norm would have been accepted in all prior cultures, or even all contemporary cultures, but it has wide acceptance in contemporary Western cultures. And if we are considering what bioethical policies to adopt in contemporary Western cultures, then it makes sense to incorporate that norm in our deliberations.

With regard to the importance of considering facts as opposed to just values in making a moral judgment, some moral judgments and proposed policies in bioethics are based, at least in part, on ignorance or misunderstanding of the facts. This is especially true in the context of the debate over stem cell research, as we will see.*

What I will do in each of the following chapters is to begin consideration of each issue, whether it is assistance in dying, conscientious objector status for healthcare professionals, embryonic stem cell research, and so on, by listing and discussing commonly accepted norms and factual beliefs that are relevant to the issue in question. These will then yield what, borrowing a phrase from the late Harvard philosopher John Rawls, I will refer to as considered judgments, that is, judgments that reflect commonly accepted norms and beliefs and in which we have a great deal of confidence.[3] This may seem too simple a method to yield practical guidance, but you may be surprised at how far we can go with premises that few will dispute. It is best not to dwell too much on the possible success of this proposed method for now; the proof will be in the pudding as we proceed to consider each issue. One virtue, by the way, of this explicit step-by-step approach to each issue will be that if there is some disagreement with one of my premises, it will be clear to the reader where that disagreement lies.

RULES OF REASONING

It would be tempting to describe the method I have outlined as a "ground-up" approach to bioethics, but that description is misleading if it assumes we can craft appropriate policies merely by reflecting on commonly accepted norms and factual beliefs. Unfortunately, arriving at appropriate policies is not quite that easy.

To begin, we may find some of our norms to be in apparent conflict. To take a very simple example, let us suppose a physician has scheduled a

---

* In urging that our moral judgments be grounded on an accurate understanding of the facts and available scientific evidence, I am not suggesting that we can deduce our values from facts. "Is" does not imply "ought." However, even though facts do not dictate our choices, they do circumscribe and shape them. Moreover, policies that are defended through use of mistaken or unsupported empirical claims are obviously flawed policies.

routine procedure with a patient, such as a mole removal. However, immediately before the scheduled procedure the physician learns that her child is involved in an accident and she cancels the appointment, at some inconvenience to the patient, so she can be with her child. Most people believe we should keep our commitments, and professionals providing specialized services may have a higher obligation in that regard. On the other hand, most people believe we also have an obligation to take care of our children and this obligation can often—but not always—take precedence over other obligations. Given the rather simple set of facts I have outlined here, I think most people would be inclined to say it is permissible for the physician to cancel the appointment, provided some sort of recompense is made to the patient, perhaps an apology combined with priority rescheduling. But we can vary the facts in a way that would suggest the opposite conclusion. What if the procedure in question is a coronary bypass for a seriously ill patient and the "accident" was the child being hit by an eraser at school? Presumably, our conclusion regarding the propriety of canceling the procedure will be different. And, of course, a more difficult decision would follow if we change the hypothetical so both situations are comparably serious, such as a coronary bypass for the patient and a car crash for the child.

The point is that we need to think about how our norms should be applied in a specific situation. This is one reason many philosophers characterize our moral norms as imposing "prima facie" obligations, that is, obligations that we must fulfill *unless* there is some overriding conflicting obligation. Most of the time it is clear what we are morally obliged to do (of course, whether we do it is another question). Nonetheless, almost anyone with any experience in life will recognize that there are situations where we have competing obligations and we need to deliberate about and justify, if only to ourselves, the choices we make.

The need to balance competing obligations shows why at least some versions of both rule-based morality and utilitarianism are inadequate. A strict deontologist who holds that morality requires following categorical rules—that is, rules with no exceptions—is going to find it very difficult to manage such situations. Either she is going to have to devise very complicated rules to govern every conceivable situation or she will find conflicts that her theory cannot resolve. The ability to appeal to consequences to determine which obligation should take precedence is one reason some find utilitarianism attractive. However, this attraction may be based on a failure to recognize one of the problems of utilitarianism in its standard variations, and that is the problem of calculating and comparing utility, however utility is defined (happiness, pleasure, desire-satisfaction, etc.). Let us say that we equate utility with happiness. Under utilitarianism, we are then required to maximize happiness, but to do that we first have to be able to calculate units of happiness. Just for oneself it is often difficult to evaluate which of two alternative actions would produce the most happiness. Will I produce more happiness for myself by buying a large house forty-five miles from work and commuting three hours a day or by buying a closet-size condominium in the heart of the city?

Trying to make interpersonal comparisons involving a range of possible actions presents significantly greater difficulties, and the notion that we can make such calculations with anything resembling precision is just fantasy. You may experience greater satisfaction or dissatisfaction from certain situations than I can appreciate. Even as simple a matter as adjusting the thermostat in one's office will produce disagreements—sometimes serious—about which temperature is the "ideal" temperature for everyone, as those who have worked in a large office environment can confirm. Decisions that potentially affect many individuals and require weighing a number of factors become even more difficult to resolve (even though some utilitarians will argue this is where their theory is most helpful). Consider issues relating to allocation of organs for transplantation. If we were thoroughgoing utilitarians, how would we establish priorities for organ transplants? Is the life of a thirty-year-old always more valuable than the life of a sixty-year-old? Does it matter if the latter is a physician and the former a toll-booth attendant? Should organ donors have preferential status for organ transplants? What if the donor has indicated a willingness to have only his corneas made available for transplant but he is in need of a liver? Does urgency of need trump time on a waiting list? Should illegal immigrants have access to organ transplants? Does it make any difference if the immigrant plans to return to his country as soon as he recovers? Of what relevance is a person's ability to pay for the transplant? Under what circumstances should we allow "directed donation" (that is, where the donor can specify the recipient of the organ)? Should this option be restricted to relatives? How close must the relative be? It would take a page or so just to list the arguably relevant criteria for crafting a thoroughly utilitarian policy on organ transplants, but that would be only the first step. One would then have to determine the utility or disutility for each individual in a class of millions for each proposed combination of criteria—when we do not even have an accepted definition of "utility." In short, utilitarianism promises us a method for making the ethically correct decision based on a calculation of the value of potential consequences, but it cannot deliver on its promise.[4]

However, it would be a mistake to conclude that because we cannot calculate which set of possible outcomes would be the best from a strictly utilitarian perspective, we therefore are incapable of making a rough weighing of harms and benefits. Almost all humans share some interests and we also have a sense of how important those interests are, at least within a given culture. We have an interest in being free from pain, maintaining our bodies in a fully functioning state, enjoying the company of friends, being free to make our own decisions about certain matters, having at our disposal property that we have acquired legally (by whatever background rules our culture has established for acquisition of property), and so on. And we have no difficulty recognizing that someone taking a bite out of my sandwich has committed a less serious harm (and perhaps no harm at all, or one readily excused) than someone taking my car. Similarly, we have no trouble recognizing that giving me a free soft drink, although a nice gesture, does not benefit me as much as contacting emergency services when I am trapped in my crushed vehicle. We make judgments such as these effortlessly and spontaneously every day. In fact, if we had to stop in every situation and think, "What should I do next?" life would grind to a halt. There is such a thing as the paralysis of analysis.

So when we do have occasion to think about competing choices, sometimes the choice is clear, even without any background ethical theory. Given our understanding of human nature and background conditions, we can balance harms and benefits promptly and can come to a decision. Of course, there will be times when the choice is not clear, because we are uncertain about how to weigh the harms and benefits, or perhaps even to make any sort of intelligible assessment of them. But the fact that it is not always clear whether an action is appropriate or inappropriate does not imply there is no difference between wrong and right or that there are no actions that are clearly wrong or clearly right. Close calls in baseball do not imply there is no such thing as a strike zone.

Accordingly, the first rule of reasoning is to weigh the harms and benefits that would result from our actions, particularly in the case where we have conflicting norms. Note that this does not imply there is one "good," whether it is pleasure, happiness, desire-satisfaction, or something else, that we must aim to maximize. I am not suggesting we employ utilitarianism to resolve moral dilemmas, but with the understanding and qualification that it is a very fallible method. In fact, another significant and more fundamental problem with utilitarianism is the notion that we can reduce all human interests to some sort of fungible, quantifiable good, and that the point of morality is to maximize this good. Persons have a plurality of interests that need to be considered, some of which are incommensurable. Moreover, if we acknowledge that we have prima facie obligations to refrain from taking certain actions, such as killing another human, and to respect the decisions of others insofar as they affect only themselves, we can see that our moral code is not easily redescribed as a vehicle for maximizing anything. Morality is a mix of norms requiring us, among other things, to benefit others, refrain from harming others, respect their individuality, distribute material goods equitably, and inculcate virtuous character in ourselves and others. Yes, morality must in some way be connected with the promotion of human interests—otherwise, what is the point? But this connection does not imply that our overriding moral obligation is to maximize any particular state of affairs.

This is an appropriate point at which to discuss our next step in reasoning. For most everyday matters, we probably do not have to proceed beyond following commonly accepted norms and, where necessary, engaging in a very rough weighing of the projected harms and benefits of following a proposed course of action. However, if we are crafting policy, we need to be a bit more systematic. In particular, we need to make sure our norms are coherent and, in the process, organize them into some set of governing principles that can both account for our current judgments and serve as a guide when we encounter novel situations—which, of course, has happened with increasing frequency in the area of bioethics.

In developing our norms into a coherent system, many moral philosophers utilize what is sometimes referred to as the method of "reflective equilibrium," also referred to as the coherence model of justification, for obvious reasons.[5] Whatever its label, the method seeks to test our initial moral judgments by detailing and examining the consequences of adhering to these judgments. One then tries to systematize the judgments and their consequences in a set of general moral principles that can explain and account for these judgments. The principles are themselves tested against our background views and theories, both moral and nonmoral. Judgments and principles that cannot be rendered consistent with one another and our background views and theories may need to be modified or discarded. Moreover, in this method, the testing and process of justification works in the other direction as well; that is, views, theories, and principles are evaluated against our considered moral judgments to determine whether or not our more general commitments require adjustments (hence the derivation of the term "reflective equilibrium"). Although not universally accepted, many bioethicists do follow this method, and it has the virtue of forcing us to examine critically many of our moral beliefs by considering their consequences and their consistency with our other beliefs.

This method also has the virtue of borrowing, with modification, a procedure successfully used in the sciences. In science, hypotheses are continually tested then modified or rejected as a result of experimental evidence. Similarly, in ethics, our moral judgments should continually be tested for adequacy by considering their practical implications. The practical implications of our moral judgments represent the "evidence" that confirms or refutes their appropriateness.*

One of the principal ways to refute a moral judgment is to show that it has implications that are inconsistent with other, more firmly held moral judgments as well as our background views and accepted practices and facts. I will use this tactic frequently in this book. For example, when I discuss assistance in dying, I will show that the claim that we can never permit assistance in dying (also known as assisted suicide) because it constitutes the intentional killing of a person is not consistent with the accepted practice of allowing competent patients to refuse life-preserving treatment with the knowledge that they will die as a result. Of course, this technique of showing inconsistencies in an opponent's judgments, beliefs, and views does not "prove" that their moral judgment is wrong. To use my example again, an opponent of assistance in dying can readjust her views so she will now also oppose the right of patients to refuse life-sustaining treatment. However, this technique does force a person to confront the implications of her views and this is at least sometimes sufficient to cause that person to rethink her views. We cannot expect anything more from any method of reasoning in ethics.

UNDERSTANDING THE RATIONALE FOR OUR NORMS

It is also important that we do not lose sight of the fact that morality is a practical enterprise with certain widely shared objectives. Moral norms help us achieve a less painful, more desirable existence by, among other things, helping to provide security to the members of a community, ameliorating harmful conditions, creating stability, allowing members of the

---

* Clearly, there are key distinctions between ethical and scientific inquiry. Among other things, as already indicated, from premises limited to factual statements, one cannot deduce an ethical conclusion. As philosophers have noted, at least from David Hume forward, one needs to employ value judgments in one's reasoning if one is going to make an evaluative conclusion. Moreover, one cannot conduct controlled studies in the moral realm in the same way one does in the laboratory. Still, there are more affinities between science and ethics than are commonly acknowledged.

community to trust and rely upon one another, and, in general, facilitating cooperation in achieving shared or complementary goals. Moral norms enable us to live together, with the benefits derived from communal activity while maintaining some discretion over our personal activities.

In making these observations, I am not suggesting that millennia ago some persons or person (or some deity) designed a code of conduct with specific objectives in mind. To the contrary, our moral norms evolved over time and without any express purposes or goals. However, we can now reflect upon and examine our norms and determine how they should be interpreted and applied to achieve socially desirable goals. Understanding and making explicit the rationale for our moral norms may justify a modification or adjustment in our norms so they can fulfill their objectives under currently prevailing conditions. Understanding the rationale for our norms can also help resolve apparent conflict among our norms.

For example, it is important to make truthful, accurate statements. If we could not rely on what others say—at least most of the time—cooperative activity would be seriously impaired. The erosion of trust in others and in institutions can be a significant problem and can result in other serious harms. However, it is easy to imagine scenarios where lying is not only permissible but morally required. The example often trotted out by ethicists is the person regarded as a loyal Nazi who is asked by a Gestapo officer if he has any knowledge of the whereabouts of a Jewish family. Unbeknownst to the Gestapo, the "loyal" Nazi is actually hiding the family in his attic. Is this person supposed to provide a truthful answer to the Gestapo? Only a morally obtuse person would think so. In this situation, telling the truth would result in immediate serious harm to innocent individuals, and there is no pressing need to worry about the erosion of trust in others that may be experienced by a Gestapo officer if and when he determines he has been deceived. We have moral rules to further human interests; human interests are not subordinate to moral rules.

I recognize the discussion so far regarding rules of reasoning and the importance of recognizing our norms' objectives has been somewhat abstract. How this methodology will be applied will become clear in subsequent chapters as I address specific policy issues. But for now, let me illustrate this method by discussing a reasonably self-contained public policy dispute in the area of medical ethics, namely, the duty of a therapist to warn someone of a threat from the therapist's patient. Ethicists will recognize that I am alluding to the situation that was addressed in the much-discussed *Tarasoff* case.[6]

In *Tarasoff*, a patient, Prosenjit Poddar, confided to a psychologist that he intended to kill a woman, Tatiana Tarasoff, after she returned from spending the summer abroad. The psychologist recommended commitment and contacted the police; the police determined Poddar was "rational" and released him; the supervising psychiatrist ordered no further action be taken. No effort was made to warn Tarasoff or her relatives. Shortly after Tarasoff's return, Poddar killed her.

The parents of Tarasoff sued, claiming that the therapists for Poddar should have warned them of Poddar's threat, and that the therapists could not excuse their conduct based on the need to protect patient confidentiality. The California Supreme Court, by a four-to-three vote, agreed with Tarasoff's parents.

We will discuss the legal implications of the ruling in the next section. Here, let us focus on the competing moral norms at issue. First, patients normally expect their communications with a therapist or physician to be kept confidential. There are good reasons to support such a norm, since proper treatment depends on accurate information. Most people regard information about their health, especially their mental or emotional health, to be of a sensitive nature and do not want that information revealed to anyone other than those professionals involved with their care. They would be less inclined to provide their therapists with accurate information if they knew this information was going to be divulged to others.

Of course, there is a competing norm here as well, namely, the obligation to assist others in avoiding serious harm especially when this can be done without endangering oneself. If I see that a bus is bearing down on you while you're standing on the corner reading a newspaper, I have an obligation at the very least to yell and warn you. Although the therapists in *Tarasoff* ran some risk if they warned Tarasoff or her parents—Poddar was obviously not averse to using violence—it is unlikely they would have been harmed had they contacted Tarsaoff or her parents.

How to resolve these competing obligations? As I have suggested, one step is to weigh the harms and benefits that follow from adhering to each norm. Possible consequences of allowing or requiring therapists to disclose threats made by patients include less effective therapy (due to incomplete information provided by patients) and this, in turn, may result in more violence by emotionally or mentally troubled individuals. Also, there would be some deterioration in the trust between patients and therapists, although once patients became aware their information might be disclosed, this harm would diminish. One cannot become disillusioned as a result of a breach of confidence when one recognizes the information one provides might be disclosed under certain circumstances.

On the other hand, not disclosing threats obviously endangers the lives of the targets of the threats, and, in most circumstances, death inflicts a serious harm. Although not every threat of violence will actually result in a violent act, and not every violent act will result in death or serious injury, the harm that could be avoided by warning potential victims does seem to outweigh the harm to the patient-therapist relationship that may result from either allowing or requiring a therapist to disclose a threat of harm to potential victims.

This tentative conclusion is bolstered by recognition that the confidentiality owed patients has never been regarded as absolute. Physicians have had the responsibility to disclose cases of certain contagious diseases as well as suspected incidents of child abuse. It would be consistent with these policies to allow or require therapists to disclose threats of serious harm; it would be inconsistent to forbid them to do so.

Finally, let us consider the underlying rationale of the norms at issue. In most circumstances, we want to preserve patient-therapist or patient-physician confidentiality because in the long run we believe this will serve individuals and society by increasing the chances that patients receive appropriate treatment. In other words, preserving patient confidentiality normally leads to improved mental or physical health and accordingly ameliorates harm. But in the situation where confidentiality would likely increase harm—by resulting in the death or serious injury of the target of a patient—it no longer makes sense to forbid disclosure. As the majority opinion in *Tarasoff* stated, "the public policy favoring protection of patient-psychotherapist communications must yield to the extent to which disclosure is essential to avert danger to others."[7] Accordingly, examining the underlying rationale of the confidentiality norm also indicates we should allow or require therapists to disclose threats made by patients.

The reader will note that throughout this discussion of *Tarasoff*, I have referred to "allowing or requiring" therapists to disclose threats. Before deciding which policy we should adopt, we also need to consider the relationship between moral norms and the law and the limits on the use of the law to shape and guide conduct. It is now time to address those issues.

THE USE OF THE LAW IN CRAFTING PUBLIC POLICY

Each day there are thousands of immoral acts totally ignored by the law. Persons lie, break promises, and engage in other inappropriate conduct, and the penalty for this conduct is the disapproval and censure of others, not a fine, a jail term, or an award of damages. Similarly, many persons fail to help others in need or distress and in most cases the law does nothing. Again, assuming this failure to help others becomes known, the punishment incurred is blame or censure; it is social in nature, not legal.

On the other hand, there are many actions that are illegal that are not immoral, except in the sense that someone might argue that any illegal act is immoral because we have a duty to obey the law. Think of the myriad regulations relating to ownership and use of vehicles, including the need for registration, vehicle inspection, restrictions on parking, and the like.

What is the connection between law and morality? Some have suggested that the law is essentially a support system for morality, enforcing the most important moral rules, while also going beyond morality in providing detailed regulations in areas where traditional morality cannot reach. There is something to be said for this view. To begin, one of the circumstances that gave rise to the development of legal systems was the need to remove severe punishment from private hands. In primitive communities, private individuals enforced norms of conduct, and this enforcement would include the infliction of severe punishment (banishment, maiming, killing), now recognized as within the domain of the law, not morality. There were at least a couple of problems with private imposition of severe punishments, including threats to the stability of the community resulting from private vendettas and the loss in efficiency from unorganized efforts at punishment of serious offenses. Also, a legal code provides for greater uniformity in punishment, which again supports the stability of the society as well as removing some uncertainty about the punishment to be expected for an offense. Moreover, it is undoubtedly true that as societies became more complex, the more complex the rules were needed to facilitate smooth, harmonious social and commercial transactions, and these complex, detailed regulations could not be readily incorporated within a set of moral norms.

However, it is misleading to conceive of the law as merely a support system for morality. The relationship is more complicated than that. Rather than thinking of the law as a support system for morality, it may be more accurate to conceive of law and morality as parallel systems of regulating conduct. There is substantial overlap between these two systems, so they often prohibit, require, or permit the same types of actions. It is also certainly true that law is informed by morality, as legal systems tend to incorporate core moral norms, and one way we criticize a law, or a public policy, is by arguing that the law is inconsistent with the community's moral values. However, law and morality also diverge, and the reasons they diverge are grounded in the limits of both systems. Having a clear understanding of these limits will assist us in crafting policy that relies, at least in part, on legal enforcement.

Morality relies on informal enforcement systems and the inculcation of a relatively small set of norms in each new generation of humans. Moral norms are effective only if they are instilled in us to the extent that they become part of the habits of thought and character of (most of) us. Accordingly, our moral norms must be relatively few in number and fairly simple because there are limits to the number and complexity of moral norms resulting not only from the costs of inculcation, but also the finite ability of humans to internalize, process, and utilize norms, given the other tasks that demand our attention. It is no accident then that, at least for most people, moral norms are usually restricted to a very discrete set (e.g., the Ten Commandments) and that norms are often phrased either as prohibitions or very broad injunctions ("do unto others," etc.). It is far easier to specify what an individual should refrain from doing than to detail the positive actions an individual is obliged to undertake. Finally, the need to keep moral norms simple explains why moral norms tend to be categorical in nature ("never lie," "never steal"), which unfortunately sometimes results in the misperception that moral norms are absolute rather than prima facie.

Enforcement of moral norms also requires awareness of the conduct of others, including the circumstances and motivations surrounding their conduct. In the small communities and groups in which humans first lived, this was obviously less of a challenge than it is today. As communities became larger, it became more difficult to punish wrongdoers and reward those who engaged in socially beneficial conduct.

I have already noted that the law provides the State with a monopoly on infliction of serious punishment. Overall, this is a beneficial development because, among other things, it provides for more uniformity and certainty. However, this comes at a cost. The State must develop a system of adjudication that is both efficient and will allow persons who do not have firsthand knowledge of the conduct to make determinations of responsibility, guilt or innocence, mitigating circumstances, and so on. This is easier said than done. Moral norms may be limited in their complexity, but legal systems are limited in the information that can be processed and brought to bear on a particular situation. Factors that we may consider relevant in a making a moral judgment about a person's conduct, such as background or the history of the relationship to us and to others affected by a person's conduct, cannot be easily considered within the confines of the legal process. A judge is expected to apply thousands of rules. She cannot memorize them, but that is not a problem because a review of the relevant code will inform her which rules to apply. What she cannot do is to spend time getting to know you as a person before applying the rules.

Earlier, reference was made to the fact that the law is indifferent to certain types of conduct considered immoral. One may think that this is because the conduct is too petty for the law to intervene. The costs of legal enforcement are certainly a factor in determining how far the law should reach, but this does not necessarily imply that conduct not addressed by the law is insignificant. Which situation is more significant for you: a co-worker who is continually deceptive and rude or someone double-parked outside your office? The law will concern itself with the latter situation, but not your work situation, principally because the latter situation can be dealt with efficiently within the legal system. This does not mean your co-worker's failings are insignificant.

Moreover, as already indicated, there are immoral failures to act that can lead to serious harm, including the death of someone in distress. Why (typically) does the law not intervene in this situation? Think of a situation where someone in the neighborhood calls for help, but not one of a hundred individuals heeds the call. Who should the legal system hold responsible? All of the one hundred individuals? If not all one hundred individuals, what factors would need to be taken into account? The age and health of the persons not responding? A determination of what exactly they perceived? (Was the cry for help clear and distinct or muffled?) What activities they were engaged in at the time? The list could go on indefinitely. Someone who is inclined to make a moral judgment about the actions and inactions of all these individuals has the option of immersing herself in the details of the situation to make a thoroughly informed judgment; the law does not have this luxury.

Moreover, with the consequences of legal sanctions being more serious, we need to consider carefully what goal would be served by imposing a legal sanction. One important reason for imposing either a criminal or civil sanction on a group of individuals would be to ensure that these individuals and others similarly situated are guided by a clear standard of conduct so they can understand their responsibilities. But how likely is it that individuals who have failed to assist someone in distress will be in the same situation again or that anyone would find himself in a similar situation? (I am fifty-four; I have never had occasion to respond to a call for help from someone outside of my family.) In addition, there is the problem of coordination of conduct among the individuals who failed to help. If any one of the hundred individuals had taken action, they all would have been relieved of responsibility. How can the imposition of legal liability for a failure to act guide me in my conduct if my ultimate liability depends on the actions of numerous other individuals with whom I may have no easy means of coordinating?

The law does impose liability for failure to act in certain situations, but it does so when the person who is held responsible has a special relationship to the individual at risk (e.g., a parent-child relationship or a physician-patient relationship). In these situations there is both recurring interaction, so it is important to define responsibilities, and a distinctive relationship, so there is no occasion to worry about coordinating with others. I do not wait to feed my child until I see what my neighbors are going to offer.

We have been considering situations where persons may act immorally, but the law imposes no sanction. However, there are also actions we may regard as morally permissible, perhaps even morally required, that we nonetheless might believe the law should prohibit, as a matter of policy. For purposes of this book, especially in connection with my analysis of assistance in dying, this is an important category of actions. Many people believe that we should not permit euthanasia or any form of assistance in dying, but, upon reflection, most of us would probably agree that at least on some occasions euthanasia is morally permissible. As an illustration of an instance of euthanasia that most would consider permissible, consider the scene in the 1992 film *The Last of the Mohicans*, in which Major Duncan Hayward is tied to a stake and is in the process of being burned alive by the Hurons.[8] Daniel Day-Lewis's character, Nathaniel Poe (Natty Bumppo in the novel), uses his skills as a marksman to put a bullet in Hayward from his position outside the Huron village, thereby killing him an instant before the flames reach Hayward. Is there anyone who thinks such an action was not morally justified? Or that s/he would not have done the same thing had s/he the same remarkable skills as Poe? Hayward is going to die in about a minute anyway; the only difference is whether he dies a horrible, agonizing death or a quick one. Nonetheless, even though we believe Poe's action was justified given the circumstances, we may also firmly believe that euthanasia should remain illegal because it would be difficult to regulate in practice and far too susceptible to abuse. When we are trying to decide whether the law should permit certain types of conduct, we need to take into account the whole range of conduct that may escape legal sanctions as a result of our decision. To take another issue in bioethics: Morally speaking, there are probably some situations in which a paid surrogacy contract would be permissible. Consider, for example, such an arrangement among acquaintances. However, it would be exceedingly difficult to craft a law that permitted paid surrogacy arrangements in the small range of cases we would find morally acceptable, while continuing to ban such arrangements in other cases.

There is no need to belabor this point. Law and morality do overlap, and each informs and influences the other, but they are distinctive systems for regulating conduct. The former places a premium on norms that can be administered by individuals with no direct knowledge of the situation in question and that will yield uniform results. The latter is more concerned with judgments made by those with direct knowledge of the situation, and uniformity of result is less important than accuracy of judgment and reinforcement of the commitment to morally appropriate conduct. Morality allows for very fine distinctions; the law is less flexible.

Let us turn now to the *Tarasoff* case once again, to apply our understanding of the law's limitations as a means of shaping and guiding conduct. The reader will recall we had decided that, at least in some cases, a therapist's duty to warn a potential victim of an attack by his patient would outweigh the therapist's obligation to maintain the patient's confidences. We were considering whether the law should allow or require a therapist to warn the potential victim.

A law allowing the therapist to warn the potential victim has the advantage of allowing the therapist the discretion to make a judgment about whether the patient's threat is credible enough to warrant a warning. This would ensure that the patient's confidences will retain a high degree of protection. It also avoids imposing liability on therapists for what essentially is a judgment call. If we were interested only in the therapist arriving at the morally appropriate decision, a statute allowing, but not requiring, the therapist to warn a potential victim might seem the best way to address this situation.

However, there are obviously other interests to take into account here, namely, the interests of the potential victim. After all, this is what first prompted us to decide a therapist should be able to reveal his patient's confidences in some instances. Allowing, but not requiring, a therapist to warn a potential victim will likely lead to underreporting of potential threats. For many therapists, their relationship with the patient will take priority. This is especially true if the therapist has no legal liability for a failure to warn absent gross neglect of professional responsibilities, which would be the implication of an "allowing" statutory provision. Moreover, if we want uniformity of conduct among therapists, requiring therapists to report any serious threat would achieve that result more reliably. Leaving it up to the therapist to decide when to report makes one wonder whether we have even created a legal standard. If some of the important functions of the law are to provide clear guidance and to standardize conduct, a statute that merely allows the therapist the option of reporting threats does not seem calculated to achieve those ends. With respect to avoiding unnecessary imposition of liability on therapists, we can achieve that goal by specifying in a statute that the therapist will be liable only if he fails to warn others of a patient's serious threat of violence. This also will give an incentive to therapists to report serious threats, since in that way they can obtain protection from a lawsuit. This may result in some overreporting, but since we have already decided that preventing harm to potential victims outweighs the need to preserve confidentiality, this result is consistent with our moral intuitions.

As it turns out, many states did enact statutes after the *Tarasoff* decision, in order to clarify the therapist's obligation. The vast majority of these statutes require a therapist to warn others of a serious threat of violence. Only two states (Florida and Mississippi) have statutes that merely allow the therapist to disclose threats.[9]

In conclusion, when deciding which policy we should adopt, we should take into account our moral norms. Moral analysis is part of formulating appropriate policies. But we cannot stop with moral analysis. We also need to determine whether our proposed policy provides guidelines that are practical and enforceable given the limitations of our legal system.

I could say much more about the relationship between morality and the law—as well as the other topics addressed in this chapter. I will expand on my remarks in subsequent chapters, as I address specific issues. However, armed now with at least an outline of the method we should use in addressing issues in bioethics, let us begin our analysis of various controversies, starting with assistance in dying.

NOTES

1. Tom L. Beauchamp, "A Defense of the Common Morality," *Kennedy Institute of Ethics Journal* 13 (2003): 260. Beauchamp and his coauthor, James F. Childress, provide a very clear exposition of common morality, including its usefulness and its limits in providing guidance on bioethical issues, in *Principles of Biomedical Ethics*, 5th ed. (New York: Oxford University Press, 2001), pp. 2–5, 401–408.

2. Bernard Gert, *Morality: Its Nature and Justification*, rev. ed. (New York: Oxford University Press, 2005), pp. 110–30, 218–19, 226–28.

3. John Rawls, *A Theory of Justice* (Cambridge, MA: Harvard University Press, 1971), pp. 19–20, 47–48.

4. Current organ distribution policy is supposed to be a blend of considerations of equity (such as time on the wait list) and utility (such as the benefit the recipient will obtain from the transplant), but "utility" in this context has been given a special definition, roughly equated to medical utility. Effectively, we have made a decision to consider only certain consequences in allocating organs, and that public policy decision was arrived at by following a balancing approach similar to what I advocate in this chapter. See the discussion of this issue in James F. Childress, "Ethics and the Allocation of Organs for Transplantation," *Kennedy Institute of Ethics Journal* 6 (1996): 397–401.

5. The method of reflective equilibrium is, of course, also borrowed from Rawls, *A Theory of Justice*, pp. 19–20, 48–50. Beauchamp and Childress (see note 1) describe how the method of reflective equilibrium can be used in the context of resolving issues in bioethics. *Principles of Biomedical Ethics*, pp. 398–99.

6. *Tarasoff v. Regents of the University of California*, 551 P.2d 334 (Cal. 1976).

7. Ibid., p. 347.

8. "Hayward's Choice," *The Last of the Mohicans*, DVD, directed by Michael Mann (1992; Beverly Hills, CA: Twentieth Century Fox, 1999). I intentionally refer to the film rather than the novel by James Fenimore Cooper because the scene I describe is not in the novel. In any event, more people have probably seen the film than read the book.

9. Claudia Kachigian and Alan R. Felthous, "Court Responses to Tarasoff Statutes," *Journal of the American Academy of Psychiatry and the Law* 32 (2004): 263–73.

3.

Legalizing Physician Assistance in Dying for the Terminally Ill

Almost everyone has some familiarity with the debates that have been raging for the last couple of decades about whether we should legalize physician assistance in dying, in particular, whether we should allow physicians to prescribe medications for terminally ill patients, which these patients can then use to end their lives. Since 1997, Oregon has authorized physician assistance in dying, pursuant to the Oregon Death with Dignity Act (ODWDA),[1] but attempts in other states to enact similar legislation have failed, although usually by narrow margins. Advocates on both sides of this issue continue to be active, and the debate over this important issue shows no signs of fading away.

I will argue that physician assistance in dying is sometimes morally permissible. In addition, there is insufficient justification to maintain legal prohibitions on assistance in dying for the terminally ill, provided physicians and their patients comply with certain regulatory safeguards.

Debates about policy issues with significant moral components often seem incapable of resolution. How can one argue with someone else's moral beliefs? However, this perception of intractability may be attributable, in part, to the way in which controversial policy issues tend to be framed. If we just shout slogans back and forth—"death with dignity" versus "sanctity of life"—we are unlikely to make much headway. Before engaging in a debate over a proposed policy, we should attempt to isolate those portions of the dispute that are capable of resolution, either through empirical evidence, conceptual clarification, or delineation of the implications of a particular position. By employing this method, we may be able to achieve a consensus. Fortunately, the debate over assistance in dying does lend itself to resolution on these terms.

One of the principal arguments against legalizing assistance in dying is that legalization will have too many harmful consequences. Persons will be put to death who really want to live and the quality of healthcare of dying patients will decline, as there will be an incentive to push them into "choosing" death. One benefit of Oregon's decade-long experience under the ODWDA is that it has provided us with the data that will determine whether these fears are justified. They are not. Indeed, there is substantial empirical evidence that legalization of assistance in dying pursuant to a statute similar to Oregon's law poses no significant threat of undesirable consequences. In fact, the evidence from Oregon's experience in the past decade should dispel many of the myths about legal assistance in dying, such as the myth that many persons will be pressured into requesting assistance in dying if it is legalized, that the "vulnerable" (usually defined as women, racial minorities, the poor, and the disabled) will be especially susceptible to pressure, and that the quality of healthcare, especially palliative care (that is, care aimed at relieving pain and suffering), will decrease, because the emphasis will be on "killing" patients rather than easing their suffering. The record in Oregon establishes that: there has been no confirmed case of any person being forced to request assistance; the overwhelming number of persons requesting assistance are white, well-educated, and financially secure, and their gender reflects that of the general population; and the quality of palliative care has actually improved in Oregon since adoption of the ODWDA.

Case closed—or a least one might think. However, opposition to legalizing assistance in dying remains strong. Part of this opposition is due to ignorance of, or a failure to give any weight to, the evidence from Oregon, but opposition also is based in part on the belief of many that assistance in dying is "killing" and "killing" is wrong; in other words, assistance in dying violates the "sanctity of life." Because the sanctity-of-life principle is sometimes characterized as a religious tenet, some may question whether it is amenable to rational discussion. But scholarly defenders of the sanctity-of-life principle usually characterize this principle as based on secular, not religious, considerations. They do so because they recognize that to be taken seriously in a public policy debate, they cannot simply claim that "my religion says this is wrong." I will show that secular arguments against legalizing assistance in dying based on the sanctity of life are not persuasive. Sanctity-of-life arguments are advanced without any understanding of this principle's underlying rationale, and they result in unacceptable inconsistencies. As we will see, one cannot oppose assistance in dying on the basis of the sanctity of life without also opposing the refusal of lifesaving treatment by competent patients.

Arguments based on the alleged harmful consequences of legalization and the sanctity of life are the principal grounds for opposition to legalizing assistance in dying. At least they are the grounds most worthy of consideration. But many other reasons have been offered to justify opposition to legalization. I will briefly consider some of these other reasons below, giving them the attention they deserve.

Whatever the ultimate explanation for the opposition to assistance in dying, it is undeniable that resistance to legalization has been effective, and thousands of terminally ill individuals needlessly suffer as a result. We owe it to them—and to ourselves and our loved ones—to subject the arguments against legalization to close scrutiny.

TERMINOLOGY

Before proceeding to my discussion, a word about terminology. I use the phrase "assistance in dying" rather than "assisted suicide." My use of this phrase does not stem from any squeamishness about suicide or a desire to gain an advantage for my argument through use of euphemisms. Instead, it reflects both a desire to be accurate in my description and an invitation to opponents of assistance in dying (or if they prefer, assisted suicide) to discuss the real issues, instead of offering arguments that score empty victories through redefinitions.

For example, Neil Gorsuch (currently a Bush-appointed federal judge) has authored a book against legalizing assistance in dying.[2] Although the book does contain some arguments that merit consideration, Gorsuch's principal argument can be summarized as follows: Intentional killing of another human is always wrong; assisted suicide involves the intentional killing of another human; what takes place in Oregon under the ODWDA is assisted suicide because the doctor intends to kill the patient; therefore, the ODWDA should be repealed and no state should permit assistance in dying.[3]

This argument has at least two questionable premises, namely, the first and the third. I will deal with the first at length below, but the third grossly mischaracterizes the physician's role under the ODWDA. Under the ODWDA, the physician consults with the patient regarding end-of-life options, and the patient decides what options he would like available, which may include access to a barbiturate that can be obtained only through a physician's prescription. Requesting the prescription only confirms the patient desires the option of hastening his death, if he ultimately deems this necessary. The patient maintains control of the process throughout and decides when, *if at all*, he will ingest the medication. Fully one-third of the patients who seek a prescription under the ODWDA *never* take the drug; others ingest the drug months after it is prescribed. To claim, in the face of these facts, that the prescribing physician always intentionally kills the patient distorts the reality of the practice in Oregon. What the physician intends to do is to ease the patient's suffering and anxiety by providing the patient with some measure of control over the timing of his death. Accordingly, use of the term "assisted suicide" can be misleading to describe the practice in Oregon.

Nonetheless, my argument is based on substance, not semantics. If the reader, for whatever reason, prefers the phrase "assisted suicide," she can continue to use that description. My arguments and policy recommendations will not be affected.

However, enough of preliminaries. Let us turn now to the reasons why assistance in dying is both morally permissible and provides an important benefit to some patients.

THE IMPORTANCE OF ASSISTANCE IN DYING FOR THE TERMINALLY ILL

Using the method I suggested in the prior chapter, I will begin by stating some background premises that reflect commonly accepted beliefs about factual matters and commonly accepted moral norms. These will yield some considered judgments about the obligation to respect a person's decision regarding matters that are important for that person, provided the decision does not place others at undue risk.

First, people have an interest in making decisions for themselves. We can call this the interest in "self-determination," or the interest in "autonomy."

Second, self-determination is both valued and valuable. In other words, most people prefer to make their own decisions, especially on important issues, so that their lives reflect their beliefs, values, and objectives. For example, at least in current Western culture, no one wants to be forced to marry someone else. As a society, we also regard self-determination as objectively valuable. We want persons to have the ability to give direction to, and find fulfillment in, their own lives. We believe that liberty and the pursuit of happiness are to be treasured and protected. Many of our laws and policies encourage individuals to take responsibility for their own lives. To cite just one example from a relevant healthcare context, the federal Patient Self-Determination Act requires hospitals and other healthcare institutions to inform patients that they have the right to specify the care they want in end-of-life situations by completing advance directives.[4]

Third, everyone in a community has to accept some limitations on his or her actions when those actions affect others adversely, including when they affect the autonomy of others. You may like to listen to country music at two a.m., but I do not want to listen to your music at that hour, maybe because I need to go to work in a few hours or maybe because I simply do not care to have your musical tastes imposed on me. Accordingly, you have an obligation to turn down your music or wear earphones.

Fourth, some decisions are more important than others because of the centrality of these decisions to a person's life. Marketing hype notwithstanding, deciding between Coke and Pepsi is a fairly trivial choice; deciding whether to have a child is not. Although there may be no clear line between critical life choices and other choices, it seems plausible to maintain that choosing a spouse, deciding whether to have a child, and deciding what career to pursue are all appropriately characterized as critical life choices. All these choices help define the identity of a person and will have tremendously significant consequences for the shape and direction of a person's life. If I am prevented from marrying the person of my choice, I have been deprived of a whole range of possibilities and I am compelled to relinquish one of the central projects of my life.

Fifth, all other things being equal, the more important a choice is for someone, the stronger the justification needed for prohibiting that person from making that choice. Note: by prohibition of a critical life choice, I do not mean ascertaining whether someone is competent to make a judgment or providing that person with relevant information that we may suspect he has overlooked. No, I mean prohibiting that person from making the choice even when we are assured he is competent to make a decision of that nature.

We can conclude from this series of premises that preventing someone from making a critical life choice results in a severe infringement of that person's autonomy and requires a compelling justification. It is not impossible to justify denying someone the right to have a child, but it would require something akin to grave and imminent danger to others.

These observations and judgments should be fairly noncontroversial. Most individuals in the United States—indeed, most individuals in Western democracies—would agree with these observations and judgments. The Supreme Court has endorsed the view that there are certain liberties so fundamental that they can be denied only under extraordinary circumstances. Fundamental liberty interests are those matters involving "the most intimate and personal choices a person may make in a lifetime, choices central to personal dignity and autonomy."[5] Among such choices are "the decision whether to bear or beget a child" and the choice of a spouse.[6]

Fine, you may say, but what does this have to do with assistance in dying?

The answer is this: there can be no more intimate, self-defining decision for a person than the decision whether to continue living. If the degree to which someone's autonomy is violated is a function of the extent to which she is forced to reorder her life, to redefine herself, then it is no exaggeration to state that a person forced to remain alive against her will has been totally deprived of her autonomy. When a terminally ill person who is competent to make decisions for herself has reached the conclusion that her life is, on balance, too painful, too degrading, too restricted to be worth living, then *for her* there are no activities worth pursuing given the conditions in which she finds herself. To force her to continue living is to condemn her to an existence that has lost any value. If a person in such circumstances longs to die but is prevented from doing so, she can no longer make any autonomous choice because her entire existence is compelled. As the legal philosopher Joseph Raz has observed, a "person whose every decision is extracted from him by coercion is not an autonomous person."[7]

But is a terminally ill person who has been denied *assistance* in dying really compelled to live? Cannot a person who is truly miserable put an end to his misery without any assistance?

In his classic 1958 article against the legalization of assisted dying, legal scholar Yale Kamisar argued, among other points, that the extent to which laws against assisted dying infringe on liberty and autonomy interests had been overstated. Prohibitions on assisted suicide, after all, are directed against the person providing assistance, not the person who receives the assistance. There are no longer any criminal penalties for suicide or attempted suicide, so, superficially, any person's interest in escaping a life he finds unacceptable could be accommodated by a simple policy of noninterference in suicides. As Kamisar put it:

> Finally, taking those who may have such a desire [to die], again I must register a strong note of skepticism that many cannot do the job themselves. . . . [A] laissez-faire approach in such matter [may be preferable to] an approach aided and sanctioned by the state.[8]

Kamisar's contention has some merit if one focuses only on the physically robust. Indeed, in most cases, offering assistance to those truly capable of doing "the job themselves" may improperly circumvent an important psychological barrier to hasty, ill-considered suicides—leaving aside any moral problems that may be raised by the offer of such assistance. Generally speaking, if a healthy, able-bodied person is too ambivalent to kill himself without assistance, suicide is for him almost surely the wrong decision. There is no need to debate here whether an act of suicide by a physically healthy individual is ever rationally defensible and morally justifiable. It seems to me that there are such circumstances. For example, consider a captured spy who might under torture reveal secrets that would endanger national security. But everyone recognizes that most suicides by physically healthy individuals are the result of either temporary or permanent emotional or mental instability. Such suicides are almost always tragedies that should be avoided, if at all possible. For healthy individuals contemplating suicide, no offer of assistance is needed or morally appropriate.

However, it is important to bear in mind that the topic of our discussion is *physician* assistance in dying for the *terminally ill*. Let us first consider the significance of terminal illness as a precondition for assistance in dying. Ending one's life without assistance is an option only for those with access to the proper means and the physical strength to use them. One crucial fact seldom acknowledged by the opponents of assistance in dying is that the State and its licensed agents control access to medications that are efficient in bringing about a peaceful death. You cannot obtain a barbiturate on your own. A physician must prescribe it and a pharmacist needs to dispense it. Because the State maintains control of the dispensation of lethal medications, a person must have access to firearms, knives, ropes, or other such means of death *and* possess the ability to use these means effectively if he is to kill himself without assistance. As the scholar and jurist Richard Posner has observed, in the last stages of a terminal illness "the patient is likely to lack the capacity to commit suicide on his own."[9] Many of the terminally ill are physically frail, and in their final weeks are confined to wheelchairs or beds.[10] For someone in such a situation, being denied assistance effectively results in that person being kept alive against his will. One reason the terminally ill have such a strong liberty interest in assistance in dying is because they need assistance to die, whereas the overwhelming majority of those who are not terminally ill require no such assistance.

Accordingly, unless we allow the terminally ill to obtain assistance in dying from a physician, we are appropriating their existence by compelling them to remain alive. The terminally ill person becomes a slave of the State, held hostage to the ideologies, dogma, and irrational fears of others.

Of course, not only does the fact that someone is terminally ill distinguish his situation from the healthy suicide in terms of respective access to the means of bringing about death, but a person's terminal illness also provides us some assurance that his request for assistance in dying is not the product of some hasty, irrational decision. Unlike the typical healthy suicide who is overreacting to some temporary setback—perhaps more imagined than real—a person who is dying is confronted with an objectively verifiable condition that will bring about his death in a relatively short time. (Under Medicare regulations and the ODWDA, a terminally ill person is someone who is expected to die within six months.)[11] The statistics collected in Oregon confirm that all those who have sought assistance have been in the terminal stages of cancer, amyotrophic lateral sclerosis, AIDS, or some other terminal, distressing condition. These are not volatile individuals reacting to serious but passing problems, such as dismissal from a job, rejection by a lover, or taunting from schoolmates. They do not seek the means to cut short a potentially long and rewarding life, but rather the means of hastening an inevitable death if and when their condition becomes truly unbearable. Simply put, their choice is between dying peacefully now and dying after pointless suffering within a few more weeks.

Perhaps not unexpectedly, some of those who oppose assistance in dying have tried to use the short period of time the terminally ill have left as a basis for dismissing the importance of assistance in dying. Essentially, the argument is, "What's the big deal? They'll be dead soon anyway."

This attitude has been exhibited from Yale Kamisar's article decades ago to Gorsuch's recent brief against assistance in dying. Kamisar asked, "Why is allowing some cancer victim to suffer a little longer too great a price to pay for the 'sanctity of life'"[12] and Gorsuch has suggested that requesting assistance in dying is "more of a lifestyle choice."[13] A "lifestyle choice"? Like taking up golf?

This condescending dismissal of the plight of the terminally ill is insensitive, to say the least. But the callousness of some opponents is not the issue. The relevant point is that even though the suffering endured by the terminally ill may be finite, it is avoidable suffering nonetheless. For some of the terminally ill, the last few weeks is a period that will be devoid of even the simplest pleasures, with a loss of functional capacity, possibly unremitting pain, and long hours contemplating the hopelessness of their condition. Unless we provide them with assistance to hasten their deaths, we are imprisoning them in a condition many of us would find intolerable. The option of requesting assistance in dying is a critically important option, and it should not be denied absent a compelling justification.

Some may question my claim that the terminally ill cannot "do the job themselves," provided they act quickly. Granted, not everyone who has just been diagnosed with a terminal illness will lack the vigor necessary to end his life through violent means. But presumably we want the terminally ill both to live as long as possible and to consult with their physicians before reaching a decision about hastening their death. These last two points are critical and are often overlooked in the debate over assistance in dying.

A ban on physician assistance in dying pushes the terminally ill to end their lives while they are still able, without discussing their plans with anyone, resulting in many precipitous or unnecessary suicides (as well as botched attempts, which simply lead to more suffering). Not only might some bring about their deaths while they have months left to live, but also others may kill themselves who, had they been able to wait to determine whether they wanted assistance in dying, would not have found their lives unbearable. In other words, assuming we desire a regulatory scheme that encourages persons to live as long as they find their lives worthwhile, a policy of permitting physician assistance in dying for the terminally ill is much more likely to accomplish this objective than banning assistance in dying. As Richard Posner has perceptively stated:

> If the only choice is suicide now and suffering later, individuals will frequently choose suicide now. If the choice is suicide now or suicide at no greater cost later, they will choose suicide later because there is always a chance that they are mistaken in believing that continued life will impose unbearable suffering or incapacity on them. . . . The possibility of physician-assisted suicide enables them to wait until they have more information before deciding whether to live or die.[14]

Let us now consider the significance of the physician's role, and why it is important not just to allow assistance in dying, but to authorize *physician* assistance in dying. The role of a treating physician is crucial in the context of end-of-life care. Legalizing physician assistance in dying allows the terminally ill to consult openly and candidly with their doctors, discussing all aspects of their condition and treatment with a knowledgeable and—one would hope—caring expert. By encouraging a frank exploration of options, a patient will be able to make an informed decision and may be persuaded to hold on longer, secure in the knowledge that he can obtain assistance when needed. Sensitive physicians will use the full extent of their training and experience in assessing the patient's condition and the prospects for effective palliative care. In consultation with the patient, they can evaluate what alternatives might be feasible and acceptable. And, of course, the physician's expertise is important for determining whether the patient is competent to make a decision concerning the course of treatment.

Failure to consider the importance of physician involvement in advising the patient regarding end-of-life options has resulted in ideologically driven and counterproductive measures, such as former attorney general John Ashcroft's effort to circumvent and gut the ODWDA by issuing a directive preventing physicians from prescribing "controlled substances" (a barbiturate is a controlled substance) to assist in a "suicide."[15] Since prescribing a controlled substance is the means by which an Oregon physician effectuates the patient's informed decision for a hastened death, enforcement of this directive would have nullified the ODWDA and put an end to legal physician assistance in dying. This was a blatant attempt to override the democratic choice of the voters of Oregon by the fiat of a Bush administration official, made all the more shameless by the federal government's disingenuous defense of the directive on the ground that it was not interfering with the decision of Oregon's voters but was merely trying to prevent illicit "drug trafficking" by confining controlled substances to legitimate medical use. Fortunately, the Supreme Court rejected the administration's position, reasoning that the attorney general was not authorized to decide on his own the legitimate boundaries of the practice of medicine.[16]

What was so shortsighted about the attorney general's maneuver was that in cutting off legal physician assistance he would not have ended assisted deaths in Oregon. He simply would have driven the practice underground, as is the case with the forty-nine other states that do not allow physician assistance in dying. Worse, some patients would have turned to assistance from nonprofessionals. In fact, as some readers may be aware, there is a fairly widespread informal network disseminating information about how patients can engage in "self-deliverance" through the use of face masks, plastic bags, and hypoxic tents, which expose the patient to inert gases and reduce the supply of oxygen to lethal levels. I do not condemn persons who turn to such methods out of desperation (or those who supply such assistance), but if we are truly interested in having a patient's decision on end-of-life choices being informed and deliberate, we will not push patients into utilizing underground methods.[17]

Legalization of physician assistance in dying certainly increases the likelihood that patients' decisions will be informed and deliberate as opposed to uninformed and hasty. And, as indicated, legalizing physician assistance in dying will prompt many to live longer than they would have otherwise.

The claim that legalizing assistance in dying actually encourages many to live longer, and some to forego hastening their death altogether, is not mere speculation. Oregon's experience confirms this. Although about 15 in 100 dying Oregonians seriously consider hastening their deaths, and although many of these discuss this option with their physicians, only about 1 in 100 decide to request assistance, only about 1 in 600 actually receive a prescription from their physician, and only about 1 in 800 use the prescribed medication to hasten death. Knowledge that escape is always available diminishes the felt need to end one's life and frank consultation with physicians can allow for a thorough consideration of other options, including alternative symptom management and hospice care.[18]

Allowing physician assistance in dying for the terminally ill shows both compassion and respect for their desire to make critical decisions about their own lives, whereas denying them the possibility of this assistance compels them to live in suffering. Moreover, legalizing this option brings added benefits, including reducing the number of premature suicides. Therefore, at least under some circumstances, assistance in dying is morally permissible, and a complete legal ban on such assistance appears unjustified. We should allow the terminally ill to have some measure of control over the timing of their deaths, absent weighty, countervailing considerations.

Of course, respect for autonomy is not a trump. Individual rights always must be balanced against the common good. It is both legally advisable and morally permissible, if not obligatory, to restrict the actions of some individuals if these actions pose a significant threat of harm to others. On some occasions, these restrictions may even require significant infringements on personal freedom, as when we quarantine individuals who have a highly infectious, potentially fatal disease. It is time now to consider in more detail the experience under Oregon's statute and how that experience refutes the one legitimate argument against legalizing assistance in dying, namely, that it cannot be done without unduly endangering or harming many patients. In other words, many patients will be coerced or manipulated into requesting assistance and the quality of healthcare for the terminally ill will deteriorate. I will also briefly address the contention that banning assistance in dying is somehow necessary to protect the patient from the consequences of mistaken diagnoses.

THE EXPERIENCE IN OREGON: INDIVIDUAL CHOICE AND EFFECTIVE PREVENTION OF ABUSE

Those who oppose legal assistance in dying cannot seem to get enough of Dr. Jack Kevorkian. He is continually invoked by them as the poster child for assistance in dying. On the very first page of Gorsuch's manifesto against assistance in dying, he refers to Kevorkian, discusses his history at some length, and then states that "Dr. Kevorkian hardly stands alone" as a proponent of assisted suicide.[19] True, there are many other individuals, including many responsible physicians, such as Dr. Timothy Quill, who advocate legalization of assistance in dying. Why then do Gorsuch and other opponents continually use Kevorkian as an example of how legal assistance in dying would operate when they know that Kevorkian's methods were *not* endorsed by any noted ethicist or responsible advocate of assistance in dying? Because, shamelessly, they want to deceive the public into thinking that Kevorkian's methods will become accepted procedure if assistance in dying is made legal. The reality is that Kevorkian's methods illustrate what happens when we maintain a legal ban on assistance in dying.

Let us discuss how Kevorkian proceeded. We will then discuss in some detail the process in Oregon, where assistance in dying is legal. Kevorkian, who by training is a pathologist, was not the treating physician for any of the patients he assisted to die. In many cases, he hardly knew them and had little knowledge of their personal or medical history. The very first person he assisted, Janet Adkins, exemplifies some of the problems in his approach. Adkins, who had Alzheimer's disease, contacted Kevorkian by phone and then flew to Michigan to meet him. After some brief discussions with Adkins over a weekend, Kevorkian hooked Adkins up to his notorious suicide-machine-in-a-van. She pressed a button and was soon dead. Kevorkian had not bothered to take her medical history, conduct a psychiatric examination (which he would not have been qualified to conduct anyway), or contact Adkins's primary physician.[20] He was not in a position to judge whether Adkins was even competent to make a request for assistance in dying. His "hurry-up-and-get-it-over-with" approach illustrates all that is wrong with assistance in dying—when it is done covertly and illegally.

That's just it. Kevorkian operated in an environment where assistance in dying was not legal. Kevorkian did not act ethically, but it is inappropriate to judge him too harshly. It is even more inappropriate to judge harshly those who sought his assistance. Kevorkian was the only available option for these individuals. Patients who were very concerned about the suffering and indignities they might have to endure in the future could not legally obtain assistance from their regular physicians, so they turned to Kevorkian.

Kevorkian at least was willing to provide his services openly (if not brazenly). Assistance in dying, indeed outright euthanasia, is practiced covertly every day in the United States. Surveys indicate that numerous physicians have acceded to their patients' requests for assistance in dying.[21] One can only hope that they have better knowledge of their patients' conditions and circumstances than Kevorkian had, but even a very conscientious physician is apt to act hurriedly in such a situation knowing that she can be prosecuted for assisted suicide or murder if her actions are discovered.

That is why the Oregon practice is so illuminating. It demonstrates that if assistance in dying is made legal, it can be provided with the assurance that the patient is competent to request assistance and all other available options have been thoroughly explored.

The ODWDA permits physicians to provide patients with assistance in dying only if a number of procedural requirements are satisfied.[22] Eligibility for assistance in dying is limited to patients whose physicians have diagnosed a terminal illness that will cause death within six months. Patients must manifest a durable, verifiable desire for assistance: The patient must make two oral requests for assistance, separated by at least fifteen days, and one written request, signed in the presence of two witnesses. Moreover, physicians are required to inform the patient of alternatives to a hastened death, such as comfort care, hospice care, and enhanced pain control. There are various other safeguards to ensure that the patient's request is informed and truly voluntary, including a confirming diagnosis by a second physician. A patient must be referred to counseling if either the prescribing or the consulting physician believes he might be suffering from a psychological disorder that can cause impaired judgment. The patient need not fill the prescription he receives from his physician, nor must he take the medication once he fills the prescription. The patient maintains control of the process throughout. If the patient decides to take the drug, he must ingest it; the physician may not administer it. Furthermore, physicians must maintain detailed records of the process leading to the prescription and these records are shared with the Oregon Department of Human Services. These records provide the basis for an annual public report that describes in detail the number of patients receiving assistance, using the prescribed medication, their medical conditions, and various other relevant data.

These records are important for assessing the dangers posed by legalization of assistance in dying. Prior to legalization, opponents of legal assistance in dying predicted thousands would seek assistance each year and physicians would place their patients on a fast track to death, many patients would be coerced into requesting assistance, and the poor, women, and minorities would be disproportionately represented among those who were coerced. Not one of these predictions proved accurate in Oregon.

Since the inception of the ODWDA a decade ago, approximately five hundred patients have received a prescription for medication that would assist them to die and about three hundred patients actually ingested it. There is no evidence that anyone who has requested assistance under the ODWDA did not have a terminal condition. There is no convincing evidence that any patient has been coerced into requesting assistance in dying. About 97 percent of those who ingested the medication have been white, and about half have been male; 41 percent had a college degree or higher, far in excess of the percentage of those with higher education in the general population. Only three of the patients, or about 1 percent, were uninsured and only seven, or about 2 percent, mentioned financial concerns as a motivation for seeking assistance in dying.[23]

In addition, there is substantial evidence that the quality of palliative care has actually improved in Oregon since implementation of the ODWDA. Again, these results are directly contrary to the dire predictions of opponents of legal assistance in dying, who argued that physicians and other healthcare workers would expend little effort to alleviate the symptoms of the terminally ill because these patients could just go ahead and die. The reality is that in a state where assistance in dying is legal, physicians and other healthcare workers are motivated to spend more time with patients in discussions about end-of-life choices, carefully exploring options and arranging for palliative care as an alternative to hastening death. Few treating physicians want their patients to choose hastened death as the first option. It is striking that fully 86 percent of the patients who have availed themselves of assistance in dying under the ODWDA were enrolled in hospice, usually considered the gold standard for palliative care.[24]

Oregon demonstrates that the problems associated with a Kevorkian-style approach to assistance in dying—hurried decisions, little interaction between patient and physician, no meaningful consideration of alternatives such as palliative care, and no assurance of the patient's competency or the strength of his desire to hasten death—are products of the ban on assistance in dying, *not* of legalizing it. If we really want to eliminate Jack Kevorkian's "practice," then let's make assistance in dying legal.

A determined opponent of legal assistance in dying may argue at this point that even if the experience in Oregon has shown that the worst feared harms and abuses have not materialized so far, there is no guarantee that they will not materialize in the future. Moreover, there is no guarantee that these abuses would not materialize in other jurisdictions were they to legalize assistance in dying.

These observations are technically correct. The past does not guarantee the future, and conditions in other states may be materially different from the conditions in Oregon. In addition, any objective observer must acknowledge the likelihood that at some point there will be at least one patient whose decision to request assistance in dying will have been the product of manipulation or coercion, by relatives, healthcare workers, or others. No regulatory system is perfect. There is a legal dictum to the effect that the abuse of a practice should not preclude its use (in Latin, *abusus non tollit usum*). However, if there were a substantial number of cases in which requests for assistance in dying had been coerced, one would have to possess an impaired moral compass not to question the wisdom of legalizing assistance in dying.

Before addressing the issue of abuse and harm to others, let me address briefly the "mistake" argument against legalization, namely, that some patients will receive a mistaken diagnosis and even though their request for assistance may be truly voluntary, they will die for no reason.

MISTAKEN DIAGNOSES

A mistaken diagnosis is a possibility, but the probability of this is vanishingly small. As indicated, there is no evidence that anyone in Oregon has received a mistaken diagnosis. Bear in mind that the diagnosis must be confirmed by two physicians.

Granted there is some evidence that a few patients who were diagnosed as terminally ill and eligible to receive assistance—that is, their deaths were projected to occur within six months—lived significantly longer than expected. In one case, a patient who requested assistance in dying lived for more than two and a half years after being diagnosed as terminally ill.[25] There is, therefore, a small possibility that some patients receiving assistance under a statute comparable to the ODWDA technically will not be eligible for assistance because their unassisted deaths would not occur within six months.

However, the unavoidable uncertainty about precisely when someone's death will actually occur does not provide a substantial reason against legalization. Recall that the patient remains in control of when to ingest the lethal medication. If the patient does not decline as rapidly as her physician predicted, the patient can simply put off taking the medication.

More important, the contention that the possibility of a mistaken diagnosis requires us to ban assistance in dying is an argument for a paternalism that is inconsistent with the respect we normally extend to self-determination. Any important choice carries with it some risk, including, for some choices, the risk of death or serious injury. Consider the risks attendant on choosing a spouse or other intimate partner. Our divorce rates demonstrate that ill-advised choices are made frequently. Moreover a mistaken choice of spouse or other intimate partner exposes a person to the risk of serious physical injury, not just emotional harm or financial ruin. Leaving aside the numerous cases of battering by spouses or partners, a mistaken choice of spouse or partner can be fatal. Roughly 10 percent of all murders annually (about 1,400) are committed by spouses, ex-spouses, or other intimates.[26] How many lives would be saved if we abridged the freedom to marry or enter into a relationship and insisted that prospective spouses and partners submit to a complete background check, psychological evaluation, and financial assessment, and also provided the State with a veto power over any proposed marriage or other intimate relationship? Presumably, we would save many from death and injury, but no one would accept such an intrusion on a decision that is so significant and personal.

More to the point, consider the latitude we allow patients with respect to decisions about refusing medical treatment. Competent patients have virtually an absolute right to decline treatment. Their decisions, of course, are supposed to be informed by a physician's assessment, but whether they are so informed or not, the patient decides what treatment, if any, to receive. Mistaken decisions are made, sometimes with fatal consequences.

Of course, we should adopt reasonable measures to reduce the risk to patients of mistaken diagnoses, but the possibility of a mistake is not an argument against assistance in dying, provided we allow competent patients to evaluate these risks and make their own decisions. Depriving everyone of a right to make a choice because some may make a mistake when they choose makes no more sense in the context of requesting assistance in dying than it does in the context of refusing medical treatment, decisions about marriage, or in any other significant decision in life.

RISKS OF ABUSE

Allowing individuals to incur risks that affect only themselves is one thing, of course, and allowing them to make choices that expose others to risks is something else entirely. We do not allow individuals to engage in target practice in public parks, however convenient such a location may be for them, and we stringently regulate the amount of alcohol people can have in their bloodstream while operating a motor vehicle principally because of the harm they may cause others. An intolerable risk of harm to those who do not want assistance in dying is a legitimate objection to legalization if in fact the level of risk is unacceptable compared with the benefits obtained through legalization. The principal harm in question is being coerced or manipulated into making a request for assistance and/or being coerced or manipulated into ingesting the lethal medication once it has been obtained. In other words, it is a risk of a hastened death that is not truly voluntary.*

How does one balance the benefits to be gained from legalization of assistance in dying against the risks to some resulting from the possible abuses of the practice? As I have argued in chapter 2, I do not believe there is a reliable way to make precise interpersonal comparisons between benefits and harms, even when conceptually there may be some relationship between the benefit and harm, as there is here (the freedom to receive assistance in dying versus a coerced or manipulated request to receive such assistance). One problem, of course, is assigning a precise "value" to these benefits and harms. I do not think that can be done. Another problem in connection with evaluating benefits and harms in this context is deciding what counts as impermissible coercion or manipulation,

---

* Another alleged harm is the possibility of a decline in the quality of palliative care, which would affect all patients. However, as already discussed, there is no evidence to support this speculation.

which obviously are vague characterizations. Is it manipulation if a spouse says, "Gee, dear, you seem to be in a lot of pain, perhaps you should discuss your options with Dr. Smith"? What if the spouse is generally a dutiful caregiver, but one day his tolerance for cleaning soiled sheets snaps and he slaps his bedridden wife? Has her request for assistance in dying two days later been coerced? These do not seem like cases of manipulation or coercion to me, but I have no doubt that some opponents of assistance in dying would so construe them.

Then we have the further problem of distinguishing between a coerced request for assistance versus a coerced ingestion of the lethal medication. The latter clearly counts as an involuntary death, but is a death involuntary if the patient was coerced into making the request, but weeks later freely decided to ingest the medication? I am confident opponents of legalization would maintain the ingestion cannot be free if the initial request was not, but this does not seem obvious. People do change their minds. (Nonetheless, for the sake of simplicity, I will hereafter simply refer to manipulated or coerced requests, on the assumption that if the request was improperly influenced the subsequent death is not truly voluntary.)

In any event, even if we were to agree on what constitutes manipulation or coercion and what constitutes a manipulated or coerced death, what percentage of requests have to be coerced or manipulated before we can justify a ban on assistance in dying? Some adamant supporters of individual autonomy might maintain that only a one-to-one ratio of coerced requests to voluntary deaths might justify a ban; on the other hand, rigid opponents of assistance in dying would argue that even one instance of a coerced request is one too many. Both positions are extreme and insupportable. If we believe that we can abridge an individual's freedom to protect others from serious harm (and this appears to be one of our commonly accepted moral norms), then if a substantial number of requests for assistance in dying are coerced or manipulated, this would argue against legalization. There is no magic number here, but, all other things being equal, a ratio of ten to fifteen coerced requests for every one hundred truly voluntary requests strikes me at least as a worrisome ratio that might require us to rethink the wisdom of legalization.[27]

However, given the fact that there is no convincing evidence of even one coerced or manipulated request for assistance in dying under the ODWDA, there is no need to spend an inordinate amount of time determining how many coerced or manipulated requests are too many. Because of the record in Oregon, it is easy to reach the conclusion that the ODWDA does more good than harm. Furthermore, given that we now have a ten-year record in Oregon, this record is sufficient to justify legalization in other states. Conceivably the situation in other states could be different, and, if it turns out that, for example, in New York, there is evidence that fifty out of five hundred requests in one year are manipulated or coerced, then legalization may be unwise in New York. But, at this stage, such a projection is simply uninformed speculation. Oregon's experience argues forcefully for legalization.

Because it is difficult to draw a definitive line between a tolerable and an intolerable percentage of coerced and manipulated requests, some might argue that we cannot determine what policy we should have on assistance in dying based on the consequences of the policy. This view just plays into the hands of extremists on both sides. ("I have an absolute right to assistance in dying" versus "Assisted suicide is always wrong.") Moreover, as explained in chapter 2, it is a view derived from faulty reasoning. Simply because we do not know exactly where to draw a line does not mean that most situations do not clearly fall on one side of the line.

In thinking about the tolerable level of abuses from legalizing assistance in dying, it is helpful to bear in mind two things: (1) a ban on legal assistance will not eliminate abuses. Some individuals will still receive assistance in dying from physicians and they will do so under circumstances where it is difficult to confirm whether their choice is truly voluntary. Indeed, if the Kevorkian example suggests anything, it suggests there may be more abuses if the ban on assistance in dying is maintained; and (2) for almost any type of choice there is always the risk that some will be harmed as a consequence of allowing individuals to make the choice instead of having the State make the choice for them. In determining what percentage of coerced or manipulated requests is tolerable, it is appropriate to consider, therefore, what level of harm to others we are willing to accept when we allow individuals to make decisions on other issues.

To reply that assistance in dying is different because here a coerced or manipulated choice results in death displays an ignorance of the consequences of some of our policies. Indeed, even with respect to relatively unimportant choices, such as the speed at which we allow individuals to drive, we tolerate a large number of deaths and other serious harms for benefits that are not very significant. Studies have shown that mandating (and enforcing) a nationwide speed limit of fifty-five miles per hour on all highways would save more than six thousand lives annually and tens of billions of dollars in healthcare costs.[28] Yet we have decided that these serious harms are tolerable so long as individuals have the freedom to drive seventy miles an hour and arrive fifteen minutes earlier in Nowheresville, Wyoming, even if their decision to drive fast results from the manipulation of passengers ("Honey, is this as fast as you can go?").

Moreover, allowing individuals the freedom to make critical life choices comparable in their significance to the decision to request assistance in dying, such as the decision to bear or beget children, *does* result in avoidable deaths and other serious harms.

For example, it turns out that if we established clear limits on who could bear children (and with implantable contraceptives this is a very feasible goal), we could save many children from death and disability. Here are the facts: Child abuse, up to and including murder of children by their parents, is a very serious problem in this county. The most comprehensive study to date on this problem determined that approximately 1,500 children die annually as the result of abuse or neglect by their parents.[29] Some experts believe the numbers of deaths are significantly underreported and that at least 2,000, and perhaps as many as 5,000 children die annually as the result of abuse and neglect.[30] Moreover, in addition to fatalities, approximately 18,000 children per year are permanently disabled by abuse or neglect. In fact, "a staggering 9.5 to 28 percent of all disabled persons in the United States may have been made so by child abuse and neglect."[31] These disabled children suffer tremendous disadvantages in life, and, of course, part of the cost of helping them overcome their disabilities is borne by the community as a whole.

Although persons of all social classes abuse and kill their children, there is a very significant, undeniable correlation between family income and child abuse. Poor couples abuse their children far more often than middle-class or upper-class couples. Government studies have demonstrated that "higher incidence rates [of abuse] were directly associated with lower income levels, and all differences among the income groups were statistically significant."[32] "Children in families with annual incomes lower than $15,000 [in 1993 dollars] had the highest rate of abuse," with their rate being "more than two and one-quarter times the rate for children in families with annual incomes of $15,000 to 29,000 . . . and nearly 14 times the rate for children in families with annual incomes of $30,000 or more."[33]

One conclusion from this impressive body of empirical evidence is inescapable: if the State were to prohibit persons from having children until they reached and maintained a certain income level, the number of murdered and seriously harmed children would plummet. Thousand of lives would be saved; tens of thousands would be protected against serious injury. A few scholars have from time to time proposed a licensing requirement for those desiring to have children.[34] Nonetheless, such proposals, although well intentioned and not completely unworthy of consideration, have almost uniformly been quickly dismissed with incredulity. Nor is there any doubt that were some legislature persuaded that restricting the rights of competent adults to bear and beget children was necessary to prevent avoidable death and serious injury to children, the Supreme Court would hold such a statute unconstitutional. "If the right of privacy means anything, it is the right of the *individual* . . . to be free from unwarranted governmental intrusion into matters so fundamentally affecting a person as the decision whether to bear or beget a child."[35] As a society we have decided that a certain amount of harm to others must be tolerated if we are to allow individuals the right to make important decisions about the shape and direction of their lives.

Choosing when to die fundamentally affects a person as much, if not more than, the decision whether to bear or beget a child. Unquestionably, we should adopt reasonable measures—as Oregon has—to ensure that as far as possible the choice of assistance in dying is truly voluntary. But demanding that we ban assistance in dying completely, because we cannot guarantee that *every* request for assistance in dying has not been improperly influenced, is starkly inconsistent with the latitude we allow individuals with respect to other critical life choices.

The fact of the matter is that persons are subject to manipulation and pressure from others at all times, but especially with respect to critical life choices such as which career to pursue, whom to marry, whether to have a child, and so on. It cannot be a sufficient justification for denying all individuals the right to make such choices that the decisions of some are the product of pressure or manipulation. Not only would such an overbearing, all-encompassing paternalism lead to a denial of the freedom to choose in all critical situations, but if the point of taking certain decisions away from individuals is to eliminate the possibility of their decision being coerced or manipulated, then such a policy would be self-defeating. Turning over all important decisions to the State substitutes the *certainty* of coercion backed up by the machinery of government versus the *possibility* of coercion from individuals trying to advance their own interests. This does not seem like an advantageous trade-off.

Finally, to drive home the point that denying the right to assistance in dying because of possible abuses is inconsistent with our other policies, we should consider an example from the context of end-of-life choices, namely, the freedom we allow competent patients to refuse treatment, even when this refusal is likely to result in death. Competent patients have virtually an absolute right to decline treatment. As one leading treatise on healthcare law states, "competent adults will have virtually no difficulty asserting their right to . . . refus[e] medical care, including life-sustaining medical treatment."[36] Unlike the requests for assistance in dying under the ODWDA, there is scant regulation of the withdrawal or withholding of life-sustaining treatment for competent patients. Relevant information, including notes regarding consultation with physicians, is supposed to be recorded on hospital charts and similar documents, and a hospital may have its ethics committee review the patient's request, but oversight of the withdrawal or withholding of treatment for competent patients is negligible.[37] Moreover, patients refusing treatment are not required to make a series of oral and written requests and to have these requests witnessed. No investigation is carried out to determine whether the patient is being improperly influenced by relatives or friends who are concerned about long-term care obligations or financial burdens. No one is required to probe the patient's reasoning to determine whether he wants treatment stopped because he finds it burdensome or because he simply wants to die.

One of the arguments often found in the literature opposing assistance in dying is the contention that patients, especially terminally ill patients, are too vulnerable and dependent on their physicians to make a truly independent, voluntary decision. This statement by Leon Kass is indicative of this argument: "To alter and influence choices, physicians and families need not be driven by base motives. . . . Well-meaning and discreet suggestions, or even unconscious changes in expression, gesture, and tone of voice, can move a dependent and suggestible patient toward a choice for death."[38] Leaving aside the point that these assertions lack empirical support and are demeaning toward the terminally ill (in Kass's view, their terminal illness has somehow rendered them infantile), this argument proves too much. If the terminally ill are not capable of making informed choices and are so easily manipulated, why then do we allow patients to refuse life-sustaining treatment? The number of patients who refuse such treatment dwarfs the couple of hundred patients who have availed themselves of assistance in dying under the ODWDA. Surely given the alleged suggestibility of these patients, there must be hundreds, if not thousands, of persons each year who are manipulated into an early death by physicians, families, or friends. If we really wanted to prevent manipulated and coerced requests that end in a patient's death, we would either deny competent patients the right to refuse treatment or, at a minimum, we would subject all patients refusing treatment to the same rigorous procedures that patients must follow under the ODWDA.

Of course, we do neither, nor, except in some extremist religious circles, is there any demand that we impose such restrictions. This inconsistency in the level of concern between assistance in dying and refusals of life-sustaining treatment cannot be explained by the risk of harm to others. As indicated, the risk of harm to others is greater in the latter set of circumstances than the former. No, as Kass admits, to his credit, this inconsistency is based on a "taboo," that is, the taboo against assistance in dying.[39] This taboo persists largely because of religious dogma. To the extent that it has anything resembling a secular justification, it is based on the illogical, unsupported claim that assistance in dying always constitutes intentional "killing," whereas withdrawal or withholding of life-sustaining treatment does not.

To sum up: When a competent person who is terminally ill decides that she no longer wishes to live in a condition she finds intolerable, allowing her to receive assistance in dying is morally permissible (arguably even morally obligatory). Denying her assistance effectively forces her to remain alive against her will and strips her of any ability to direct her own life. What limited choices she has left (for example, whether to have orange juice or apple juice while she is lying in her hospital bed) are all compelled choices because she no longer desires to engage in *any* activity. When the State both controls access to medications that this person could use to end her suffering and also legally prohibits physicians and others from providing effective assistance to this person, then the State has appropriated her life. This represents a substantial infringement, if not a total denial, of this person's liberty. Under some circumstances, the State might be able to justify such a substantial interference if respecting this person's liberty interest would cause a significant risk of imminent, serious harm to many others. However, in the context of assistance in dying, as demonstrated by Oregon's experience, legalizing assistance in dying does not pose an intolerable risk of harm to others. In fact, the evidence indicates that with appropriate regulation assistance in dying does not pose even a negligible risk of harm to others. Furthermore, when we compare assistance in dying with other practices, we find that we are willing to accept far greater risks of harm for the sake of individual liberty.

Accordingly, we should legalize assistance in dying, with procedures similar to those used in Oregon, in the rest of the country.

As I have acknowledged, opposing legalization of assistance in dying because it would cause many persons to be put to death involuntarily, through coerced or manipulated requests for assistance, is a legitimate argument. There is no evidence to support this argument, however. In addition, opponents of legalization have put forward several other arguments. These arguments are based largely on bad reasoning of one sort or another. It is now time to address these arguments, beginning with the argument based on the so-called "sanctity-of-life" principle.

OPPOSITION TO ASSISTANCE IN DYING BASED ON THE SANCTITY OF LIFE

There is little doubt that much of the opposition to legalizing assistance in dying derives not from concerns about possible abuses of the practice, or the risk of harm to others, but from the conviction that assisting someone to die is itself an inherently immoral act. This conviction is in turn based on the sanctity-of-life (or, as it is sometimes called, the inviolability-of-life) principle. In arguing against assistance in dying, the late Pope John Paul II stated, "The deliberate decision to deprive an innocent human being of life is always morally evil and can never be licit either as an end in itself or as a means to a good end."[40]

But what does this verbiage mean exactly? For example, what does it mean "to deprive" someone of life? Can one "deprive" a person of life when that person expressly requests—even pleads—for an end to suffering? Do we "deprive" someone of pain when we provide them with an analgesic at their request? And what does it mean for something to be "an end in itself or as a means to a good end"?

As we will see, the sanctity-of-life principle is less a carefully articulated moral view than a welter of confused and contradictory attitudes. In particular, I will show there are two principal flaws in the claim that we cannot permit assistance in dying because it violates the sanctity of life.

First, it is not possible to draw a distinction between assistance in dying (considered impermissible) and refusals of treatment (considered permissible) on the ground that the former always constitutes an intentional killing, nor does such a distinction make any sense from a policy perspective. Moreover, banning assistance in dying fails to promote the underlying rationale of our norms against killing others. We condemn killing because killing is almost always harmful to the interests of others. Hastening the death of a terminally ill person may not be harmful to that person's interests, however; certainly if the person requests such assistance there is a presumption that this assistance furthers his interests. As I explained in chapter 2, in applying our moral norms sensibly, it is critical to understand the rationale and objectives of these norms.

CONFUSION ABOUT CAUSATION AND INTENTION

Let us begin our analysis of the arguments derived from the sanctity-of-life principle by examining the allegedly profound moral distinction between the withdrawal and withholding of life-sustaining treatment and assistance in dying. Those who reject physician assistance in hastening death often attempt to distinguish between permissible cessation of treatment and what they characterize as the immoral act of assisting in a suicide. They attempt further to ground this distinction in what they describe as the difference between "letting die" and "killing." Withdrawals or withholdings of treatment have usually been classified in the "letting die" category. This distinction between letting die and killing has long been the most critical one in attempts by opponents of assistance in dying to distinguish appropriate from inappropriate means to death.

In fact, this distinction between letting die and killing formed the principal basis for the Supreme Court's companion decisions in *Washington v. Glucksberg* and *Vacco v. Quill*, the two 1997 cases in which the high court ruled that patients did not have a constitutional right to assistance in dying.[41] In so ruling, the Court unambiguously stated that while "*Everyone* . . . is entitled, if competent, to refuse unwanted lifesaving medical treatment; *no one* is permitted to assist a suicide."[42] Note that in describing the legality of cessation of treatment, the Court focused on authorization by the patient. All competent patients may insist the physician stop treatment, even if it results in death. Nonetheless, when the Court distinguished assistance in dying, it focused not on the patient's right to make a decision about his treatment, but on vague notions of causation and intent.

Thus, according to the Court, one key difference between assisted dying and refusing lifesaving medical treatment is that "a patient refus[ing] life sustaining medical treatment dies from an underlying fatal disease or pathology," not the physician's actions. The Court also employed that obscuring word of last resort, characterizing a death from a refusal of treatment as "natural."[43] With respect to intent, the Court stated that when a physician withholds or withdraws treatment, the physician "purposefully intends, or may so intend, only to respect the patient's wishes," whereas a physician "who assists a suicide, however, 'must, necessarily and indubitably, intend primarily that the patient be made dead.'"[44] Similarly, the Court concluded that a patient seeking assistance in dying "necessarily has the specific intent to end his or her own life, while a patient who refuses or discontinues treatment might not."[45] Therefore, according to the Court, both in terms of the cause of death and the intentions of the physician and patient, cessation of treatment can be sharply distinguished from assistance in dying.

With all due respect to the Supreme Court, this is just hogwash. These distinctions are totally untenable. About the only favorable thing that can be said about the Court's analysis is that it mercifully spared us the even more specious distinction between omission and act, which some scholars and prior court decisions had seized upon as the key distinction between cessation of treatment and assistance in dying. The argument for this distinction is essentially this: cessation of treatment is merely omitting something, whereas providing assistance in dying is an action. This is a mere semantic distinction at best. To begin, "omissions" can cause harm as much as "actions." (If I stand by and calmly watch you choke on some food when I could have applied the Heimlich maneuver or called for help, my "omission" has certainly contributed to your death.) Moreover, contrast a physician who yanks a ventilator from a patient, who then dies within a couple of minutes, and a physician who writes a prescription that the patient may use—if at all—weeks later. Which of these is best described as an "action" directly connected with the patient's death? Stopping treatment is clearly as much of an action as writing a prescription.

The foregoing rhetorical question also helps to illustrate why the causation distinction invoked by the Supreme Court is faulty. To say that an act (or omission) is the cause of a patient's death is to reason backward from our conclusions about the proper assignment of fault or responsibility. If a neighbor's child falls in my swimming pool and drowns, did I help cause the child's death? The law may, in fact, hold me liable if I was aware that children are in the neighborhood and made no effort to inhibit access to my pool. What if I build a fence around my yard, but a child climbs the fence anyway? I probably would not be held liable, because the law will reward me for taking precautionary measures, but the relevant sequence of events is the same. I build a pool; child is attracted to the pool; child falls in the pool; child drowns. In other words, "causation" in the law is a readily manipulated concept that is used to advance policy objectives. We want swimming pool owners to take precautions, so we relieve them of causal liability when they take those precautions. We want physicians to respect patients' desires concerning treatment, so we say that the physician's removal of the respirator does not cause the patient's death. But, of course, given the patient's condition, the removal of the respirator could be characterized as both a necessary and a sufficient cause of the patient's death. The patient would not have died at that time but for the physician's removal of the respirator.

By contrast, under the ODWDA, a physician who prescribes a lethal medication at a patient's request is simply writing a prescription. That act no more "kills" a patient than does the writing of a prescription for sedatives or analgesics for a patient who is undergoing withdrawal of treatment. Under the ODWDA, the patient must make a conscious decision to use the drug. As indicated previously, about one-third of the patients who obtain a prescription under the ODWDA never ingest the lethal drug; others ingest it months after it has been prescribed. For those who do take the drug, the physician's writing of the prescription is a necessary step in the process that leads to the patient's death, but it is not the determinative or even the final step. Given this background, to say that the physician who assists a patient's death under the ODWDA causes the patient's death, whereas a physician who removes life support does not cause a patient's death, is simply another way to say that we should not allow the former type of conduct whereas we should allow the latter type of conduct. Significantly, even some scholars such as Gorsuch who strongly oppose legalization of assistance in dying concede that "causation may be an imperfect basis" for distinguishing between cessation of life-sustaining treatment and assistance in dying.[46]

Gorsuch does endorse the use of intention to distinguish between cessation of treatment and assistance in dying. Gorsuch argues that assistance in dying (which, of course, he refers to as assisted suicide) "*always* involves, on the part of the principal, an intent to kill and also requires that the assistant intentionally participate in a scheme to end life,"[47] whereas cessation of life-sustaining treatment need not involve an intent to end life. This argument is based on the observation that patients refusing treatment may simply want to be free of the burdens of medical care or, as Gorsuch puts it, "they are tired of the invasive treatments and tubes and the poking and prodding that have come to characterize much of the modern medical care" and they may wish to maintain their sense of dignity and privacy and die peacefully at home with their loved ones.[48]

It is undoubtedly true that many patients refuse treatment for precisely the reasons Gorsuch notes. However, they do this with the understanding that the cessation of treatment will almost surely result in their death. Therefore, their situation is not in any way distinct from the terminally ill patient who, for similar reasons—desire to avoid burdensome treatment and suffering, maintain dignity, die peacefully with loved ones—takes a medication with the understanding that the medication will almost surely result in her death. What is the difference here? Those who argue for a distinction usually emphasize that the patient who refuses treatment does not necessarily want to die, but merely wants to be freed from the degrading intolerable condition in which she finds herself. Were she to recover after treatment was stopped, she would not be disappointed. However, exactly the same can be said for the patient who ingests a medication that is likely to cause death. If through some improbable and unanticipated chemical reaction the barbiturate did not result in death but rather in a remission of the cancer or other terminal condition, the patient would be overjoyed, not disappointed. If we determine intent not by examining the patient's understanding of what is likely to happen, but rather the motivations and desire of the patient, there is nothing to distinguish the terminally ill patient who dies as a result of stopping treatment and the terminally ill patient who dies as a result of ingesting a barbiturate. Both patients are knowingly taking actions that almost certainly will result in their deaths and their motivations and desire may be, and probably are, identical.

The only reason there is a superficial appeal to the distinction based on intention is that those who employ this distinction covertly shift their focus when they move from discussing cessation of treatment to assistance in dying. In discussing cessation of the treatment, they focus *not* on what the patient knows will happen, but rather on the patient's motivations and desires. Then, when discussing assisted dying, the focus switches to what the patient knows will happen and the patient's motivations and desires are ignored. This is nothing more than a shell game—sadly with suffering patients as the victims.

As the foregoing argument demonstrates, distinctions based on causation and intention fail to distinguish cessation of life-sustaining treatment, which is considered morally and legally permissible, and assistance in dying, which is condemned by many and legal only in Oregon. These distinctions are unsatisfactory for a number of reasons, not least because they tend to mask, rather than promote, consideration of the relevant factors that ought to be considered in determining permissible conduct. The key consideration in determining the permissibility of both cessation of treatment and assistance in dying is whether the decision of the patient is informed and voluntary. Withholding or withdrawing treatment from a competent patient is morally and legally justifiable if the patient has made an informed, voluntary decision authorizing the cessation of treatment. The physician is not the relevant cause of death and does not act wrongly if he or she has valid authorization for withholding or withdrawing treatment. By contrast, if a physician removes a respirator from a patient who needs it and wants to continue to use it, the action is wrong, even though the physician has only removed artificial life support and let "nature" take its course. Absent the patient's authorization, such "letting die" is simply killing. The lack of authorization by the patient is the relevant consideration in determining the act is unacceptable.

Similarly, when a patient requests a physician to assist him in hastening his death by facilitating access to a barbiturate, the key consideration is whether the physician's action is authorized by an informed, voluntary decision of the patient. If that is the case, then there is no sound reason for thinking the physician's or the patient's action is morally impermissible. The physician merely provides the patient with the means to escape the ravages of a fatal illness, exactly as the physician does when he disconnects life support. Furthermore, there is no legal justification for distinguishing between the two types of actions, assuming there is no difference in the risk of harm posed to others by these actions (and we have shown there is no such difference).

Why then do scholars, jurists, and physicians who clearly are not unintelligent persist in arguing for a distinction? To the extent they are not mistakenly swayed by the argument that legalization of assistance in dying will result in too many abuses, they are motivated, I submit, by the view that assistance in dying is somehow inherently wrong. In other words, such conduct violates the sanctity-of-life principle. Logically, this should lead them also to oppose cessations of life-sustaining treatment, for the reasons stated, but most of us cannot accept the notion that somehow we are permitted to force a ventilator down a resisting patient's throat in the name of preserving the sanctity of life. There is an instinctive revulsion to the idea of restraining a patient to hook her up to machines, even if doing so is necessary to keep her alive. (This revulsion is reflected in tort law, which has long made unwanted "touchings" actionable; you have a personal space that I cannot invade absent your consent.) The same instinctive revulsion is absent from the situation in which a patient is not forcibly attached to machines, but is merely dying a slow, painful death from cancer or some other terminal condition. In the latter situation, because we are not required to take "hands-on" action to compel the patient to live, some—especially physicians and nurses, who are the ones who interact directly with patients—are more comfortable with insisting on maintaining the sanctity of life. They can act from a distance, as it were, by keeping in place laws that prevent patients from obtaining medications that would allow them to die in peace. They are not required to do something as unseemly as strap a struggling patient down to insert an IV line. Moral distance can produce different moral conclusions, even though logic dictates identical results.

But now we need to dig deeper. We have shown that it is inconsistent to object to assistance in dying because of the sanctity-of-life principle while raising no objection to cessation of life-sustaining treatment. Nonetheless, there still will be some who maintain that there is a significant difference in the patient's intent in the two situations. To complete our argument, we need to take a close look at the sanctity-of-life principle to determine whether it has a sound basis or whether it is a principle that represents a misunderstanding of the rationale for our prohibition of killing.

WHY DO WE PROHIBIT KILLING?

The sanctity-of-life principle holds that intentionally bringing about the death of a human being (including oneself) is always morally impermissible. Its defenders see it as independent of any other moral norm; in other words, bringing about the death of a human being is wrong regardless of whether this death produces bad consequences, violates a special obligation toward that person, and so on. Usually exceptions are grafted on to the principle so that the principle applies only to "innocent" human beings (thereby allowing killing of criminals, combatants, etc.), but that is a complication we need not worry about for our purposes.

The principle has some intuitive appeal, but we need to ask ourselves, *why* do we prohibit bringing about the death of others? If morality has a point, and it is not simply some ingrained set of instinctive reactions that serve no purpose, our moral rules and norms must have objectives that are consistent with the manner in which we want to structure our interpersonal relations. As discussed in chapter 2, I believe there are certain core moral norms that one can find across cultures; one such norm is the rule prohibiting bringing about the death of another person, a.k.a. killing. The cross-cultural commonality of this rule suggests that it is critical for people to live together peacefully. A moment's reflection will show why the prohibition on killing serves the needs of the moral community. Killing harms others. If we are to maintain peace within the moral community, we cannot allow people to injure, thwart, or defeat the interests of others without appropriate justification, and physically harming a person can be an effective way of injuring or thwarting another's interests. Killing that person is usually *the* most effective way. The person who is killed has had his interests injured with extreme prejudice. His interests have been terminated.

All of this should be fairly straightforward. Only the morally deranged believe they are permitted to kill with impunity, and we all recognize that killing is normally the most serious harm we can inflict on another, so we condemn killing morally and punish it legally.

But notice that despite our normal condemnation of killing and the infliction of other harms, we tolerate and even applaud the infliction of harm in some cases (even beyond the special situations of combat or criminal executions, which can be ignored for purposes of this discussion). Moreover, the principal reason we tolerate the infliction of harm in some cases is that a competent person upon whom the harm has been inflicted has authorized the imposition of harm. This consensual infliction of physical harm takes place countless times each day—for instance, in doctors' offices, dentists' chairs, hospitals, gymnasiums, and similar settings. Persons authorize harm to themselves because they want to accomplish certain ends that cannot be achieved except via the infliction of this harm, whether it is the removal of a tumor, a pretty smile, slim body, and so on. Slicing open someone's abdomen is normally a grievous wrong; it ceases to be a wrong when the knife is wielded by a surgeon and the surgery is authorized by a competent patient.

Can the same reasoning be applied in the case of hastening someone's death? Yes. In normal circumstances bringing about someone's death is a moral wrong because the death completely deprives the person of any future ability to pursue and fulfill his interests. But one who seeks to be put to death because he wishes to avoid existing in a wretched, intolerable condition will not be deprived of any desired future state because he has no interest in remaining alive given his condition. Accordingly, just as we recognize an exception to the general rule of "inflict no harm" in situations in which the harm, in conjunction with other conditions, does not, on balance, impair a person's interests, similarly we can recognize such an exception in the case where the harm in question is death, but, on balance, this harm is in the person's interest because it delivers the person from an intolerable condition.

Granted, just as with any serious harm, there should be safeguards that ensure, among other things, that the person requesting the harm is competent to make a decision, that the person who assists in bringing about the harm has the appropriate knowledge and skill to counsel the person, discuss alternatives, and effect the harm in a way calculated to achieve the person's objectives, and that background conditions establish that the person's request to be harmed is rational and not a momentary whim. These safeguards are important because harms are not normally reversible, at least not easily. A dentist will not remove a patient's healthy teeth merely because the patient walks in and asks that this be done. Likewise, no physician is going to assist a patient to die merely because he walks into the physician's office and requests a barbiturate. When a patient is terminally ill, however, background conditions indicate that the person's request for assistance in dying is not necessarily whimsical or irrational. Moreover, if the patient's condition indicates the patient is suffering and the patient considers this suffering to be intolerable, then hastening death can be a rational objective. For the terminally ill then, we should recognize an exception to the general rule that bringing about someone's death is morally impermissible.

The argument for this exception to the general rule is especially strong because it actually supports the general prohibition. The "exception" is implied by the general prohibition once we understand the reasons for the prohibition. Furthermore, the argument also shows why intuitively we may feel that bringing about someone's death is an especially grave harm. We could set forth the scheme for this argument as follows:

1. All other things being equal, one ought not to harm a person's interests.
2. Physical damage to a person will almost always harm a person's interests.
3. Physical damage to a person may be justified if it is consistent with that person's interests as determined by his own competent evaluation and statement of those interests and other background conditions.
4. Death is an extreme form of physical damage, as it forever prevents a person from pursuing his interests.
5. Accordingly, bringing about someone's death is almost always unjustified (conclusion from 1, 2, and 4).
6. However, bringing about someone's death may be permissible if it is determined from that person's own competent evaluation and statement of his interests and other background conditions that death is consistent with his interests (conclusion from 1 and 3).
7. Nonetheless, given the extreme, irreversible consequences of death, bringing about someone's death is likely to be justified only when that person's continued existence is severely impairing his interests as determined by his own competent evaluation and statement of his interests and other background conditions (conclusion from 1, 3, and 4).
8. It is permissible to assist in hastening the death of a terminally ill person who is suffering and is competent to assess his condition and insists that, given his condition, he has no interest in remaining alive (conclusion from 1, 3, 4, and 7).

In other words, as is true with any moral norm, the general prohibition against bringing about the death of another should not be considered a rigid taboo that we adhere to mindlessly. We need to understand the norm's rationale and the objectives we are trying to achieve by inculcating and following such a norm. Ultimately, this norm, and many others, serve to respect, protect, and further the interests of others. It is almost always morally impermissible to inflict the harm of death because it is prejudicial to the interests of others. But in the exceptional case, where the death in question merely hastens an already inevitable death and allows someone to escape the ravages of her suffering, death is not necessarily prejudicial to that person's interests. To insist otherwise is to adhere unthinkingly to dogma; it is to follow a rule without any understanding of why we have the rule. But, essentially, that is what the defenders of the sanctity-of-life principle do. They maintain that intentionally causing the death of another is wrong, period. It does not matter whether this principle serves any function related to human interests.

Before moving on to consider an argument in favor of the sanctity-of-life principle that has some superficial appeal, I will briefly discuss and dispose of a meritless contention that one finds all too often in the discussions concerning assistance in dying. This is the contention that assistance in dying necessarily harms a person's interests because the person will have no interests after death; they will be eliminated. The person who hastens his death will not be pain-free because he will not be around to experience a pain-free condition.[49]

This is mere wordplay. People throughout history have accepted the notion that there are situations worse than death. Unfortunately, some persons find themselves in situations that are utterly wretched, degrading, and unbearable and from which the only escape is death. There is no mystery why persons in such situations will find death to be something in their interest even while recognizing that death will put an end to their existence. Existence can have negative value. Patrick Henry was not in any great distress or torment when he uttered the phrase "give me liberty or give me death,"[50] and some today may not endorse his sentiment, but still his sentiment was rational and intelligible even though he obviously would have no freedom of action after death. For him, not existing was not as bad as existing under what he regarded as a dictatorship. Similarly, for some of the terminally ill, not existing is not as bad as existing in a condition in which one can do nothing but suffer. It is not that the terminally ill patient will experience an improvement in his condition after his death; it is that he will no longer have to experience nothing but misery. Zero is greater than any negative number.

THE BEST ARGUMENT FOR THE SANCTITY-OF-LIFE PRINCIPLE

Let us now turn to examining the best argument in favor of the sanctity-of-life principle. Some proponents of assisted dying might claim this is unnecessary because the sanctity-of-life principle is a religious viewpoint, not a secular one. Historically, it is true that the sanctity-of-life principle has rested on certain theological assumptions, including the belief that God has some ownership interest in our lives, and in bringing about a death, including our own death, we act against the Deity's interests. The very use of the term "sanctity," which implies that life is sacred, shows the religious roots of this principle. But many who defend the sanctity-of-life principle have attempted to ground this principle on secular concerns and their arguments deserve consideration. Moreover, proponents of this principle represent one of the most powerful forces blocking the acceptance of assistance in dying. To disregard their arguments diminishes the practical value of any discussion of assistance in dying.

I am going to discuss the version of the argument advanced by Joseph Boyle.[51] It is similar to versions offered by Gorsuch and others,[52] but I believe Boyle's argument is both more philosophically sophisticated and clearer. Boyle views the sanctity-of-life principle as codifying in practical terms the implication of recognizing human life as a "basic good" of human nature. Boyle maintains that a justification of the sanctity-of-life principle, and its application to assistance in dying (which he groups under the category "self-killings"), may be summarized as follows:

1. One should never act with the intention of destroying an instance of a basic good of human nature.
2. Human life is a basic good of human nature.
3. One should never act with the intention of destroying an instance of human life.
4. Intentional self-killing is acting with the intention of destroying an instance of human life.
5. One should never act with the intention of killing oneself (which, again for Boyle, implies that one cannot hasten one's death, with or without assistance).[53]

Premises 1 and 2 are the critical premises here. I am going to concentrate on Premise 2, but I do want to address Premise 1 briefly.

Premise 1 assumes, among other things, not only that there are certain "basic goods" of human nature, but also that one can always act in such a way that no moral dilemma results, that is, one never has to choose among alternative courses of action, each of which entails the intentional destroying of a basic good. This is a controversial premise. Philosophers will recognize that for Premise 1 to work, indeed for the sanctity-of-life argument to get off the ground, Boyle and others who accept this argument need something called the Doctrine of Double Effect (DDE). Stated as simply as possible, DDE holds that an action that has a foreseen bad effect, as well as a good effect, may nonetheless be permissible if four conditions are met: (1) the action itself is good or morally neutral; (2) the person performing the action intends only the good effect; (3) the bad effect is not a means to the good effect; and (4) the good effect has positive consequences that, in some sense, outweigh the bad effect.[54] Defending DDE requires, at a minimum, acceptance of a controversial understanding of intention. For example, we would have to accept Gorsuch's viewpoint that patients who stop treatment with the knowledge that they will die do not intend their deaths, whereas patients who hasten their deaths under the ODWDA do intend their deaths—even though both groups of patients may have the same motivations and desires (for example, to be free of suffering, to die with dignity, not to burden their families) and are equally certain their actions will result in their deaths. We should reject that malleable notion of intent, which in many cases simply provides a way to justify what one approves of, and condemn what one disapproves of, through self-serving descriptions of the events in question.

Consider a couple of examples, one biblical, the other contemporary. The Old Testament relates the story of Samson, informing us that he brought a temple down upon himself and the Philistines who had been tormenting him, asking God both for vengeance and death for himself.[55] Did Samson commit suicide when he brought the temple down? Certainly St. Augustine thought so,[56] but if one were so inclined one could describe his death as the unfortunate side effect of inflicting appropriate punishment on the Philistines.[57] Same action, two different descriptions, two different ethical conclusions (for those who believe suicide is immoral). Now consider terminal sedation, the practice of using increasingly heavy doses of sedation (opioids, benzodiazepines, and barbiturates) to induce a stupor in a patient with intolerable pain. This is usually combined with withdrawal of nutrition and hydration. The patient typically dies within days. Opponents of physician assistance in dying often suggest that it is unnecessary because terminal sedation is available as an alternative and terminal sedation does not pose the same moral problems because the physician does not "intend" the patient's death. Oh, really? It seems to me that terminal sedation is just euthanasia without the safeguards of assistance in dying, a view shared by many others. As David Orentlicher has noted, "Although death from dehydration or starvation during terminal sedation resembles death resulting from the withdrawal of treatment, it is in principle more like euthanasia."[58] I am not going to pursue an extended discussion of DDE. Many opponents of the sanctity-of-life principle already have expended a considerable amount of energy attacking DDE, viewing it, not unreasonably, as a significant weakness in arguments in favor of the sanctity-of-life principle, and there is no point in my replicating their efforts. The interested reader can review some of the vast literature on this topic.[59] The main point I want to make here is that any ethical principle so readily manipulated to obtain a desired result is practically worthless as guide for conduct.

The assumption in Premise 1 that there are certain "basic goods" of human nature is also not beyond dispute. Nonetheless, the notion that there are some goods for human beings with which morality should be concerned is fairly widely accepted. As indicated, it is part of our core morality that, all other things being equal, one should avoid harming a person's interests. It is true that "interests" and "basic goods" do not have an identical meaning nor do they necessarily refer to the same things. As we will see, Boyle uses the term "basic good" to designate something that can be intrinsically valuable. The term "interests" does not have these implications. Still, there are probably some interests that almost all human beings share. For example, one would be hard-put to think of a situation in which preserving one's mental competence is not in one's interest. Mental competence and rationality are plausible candidates for goods that, if not universal in nature, are so near to being so that this distinction is of little importance. Provisionally, we will accept Boyle's concept of a basic good, although we will examine it more closely when considering Premise 2.

I want to focus on Premise 2 because Premise 2 is really what makes the sanctity-of-life principle distinctive, as Boyle himself points out.[60] If this premise is removed, then one simply has a rule-based ethic condemning killings, including self-killings. No one is going to be persuaded by the mere pronouncement of such a rule.

What then of Premise 2? The key to Boyle's defense of human life as a basic good is his assumption that we can divide goods into basic and "instrumental" ones and his contention that we regard life as a basic or fundamental good. (Boyle uses the terms "basic" and "fundamental" interchangeably.) Instrumental goods are things that are valued only as a means to an end. Sun block is an instrumental good. We do not purchase sun block lotions or creams to display on our bookcases nor do we shape our lives to maximize the acquisition of sun block; sun block can be very valuable, but only as means to protecting our skin. According to Boyle, life must be a basic good because it "certainly is not extrinsic and instrumental."[61] That is to say that we do not value life only because it is a useful means for obtaining other goods. Moreover, because it is not an "extrinsic" good, it has value independent of its relation to other goods. In other words, preservation of life is justified by the preservation of life and nothing further. We do not value life for the consequences it produces, but rather it is desired for its own sake. (Gorsuch similarly observes, "To claim that human life qualifies as a basic good is to claim that its value is not instrumental, not dependent on any other condition or reason, but something intrinsically good in and of itself.")[62] In sum, Boyle maintains that goods can be divided into two exhaustive categories, namely, instrumental and basic goods, and that this division, in Boyle's mind, correlates with the distinction between goods that have extrinsic value (that is, their value depends on their relationship with other things) and goods that have intrinsic value (that is, their value depends on nothing else). Life must be a basic good with intrinsic value because it clearly is not a merely instrumental good.

So far, Boyle's argument has some plausibility; at least it seems to be relying on a distinction often made between things valued for their own sake and things valued only as a means to some other good. Moreover, if our only choice is to classify life as something valued for its own sake or as something valued only for the sake of something else (that is, an instrumental good), Boyle appears to have a point. It does seem implausible to regard life as an instrumental good in the same way that sun block or money is an instrumental good. Life does not so much produce other goods as it is a necessary constituent of whatever goods we achieve.

However, Boyle's argument has plausibility *only* if we accept that all goods must be placed in one of two categories: intrinsically valuable goods and instrumentally valuable goods. This is the critical flaw in his argument. If we introduce a category of goods that are extrinsically valuable, that is, dependent for their value on other goods, but that can nonetheless serve as an end or goal of human action (that is, they are not valued only as a means to something else), then Boyle's argument collapses.

Human life seems to be such a good, that is, one with its value dependent on its relationship to other goods. Admittedly, human life does not seem a mere instrumental good; human life can convincingly be described as something that is valued as an end or goal. Nonetheless, the value of human life is dependent on its relations to other goods, for example, rationality, cognition, sentience, and when these relations are severed, the value of human life is substantially diminished, if not eliminated. That the value of life is extrinsic, not intrinsic, is supported by the consideration that no one regards living in a persistent vegetative state to be desirable. Yet if we strip away all goods except life itself, all that would remain would be a bare, biologically human existence, devoid of cognition, emotion, or any experience. Bare, biologically human life does not appear to be a good "desired for its own sake." Life's value is dependent on its relationship to other goods.

The view that we cannot exhaustively divide all goods into intrinsically valuable and instrumentally valuable goods, that is, we can have goods that we value as an end or goal but that nonetheless depend on their value for their relationship to other goods, is not original with me. A number of philosophers have made similar observations. For example, in an insightful essay Christine M. Korsgaard has argued that the distinction between final and instrumental goods does not always match the distinction between intrinsic and extrinsic goods.[63] One can value something as an end or goal that has value only in relation to other goods. This does not imply that the thing in question is only a means to an end; it implies the thing in question has value in a certain context. Likewise, to say that something has intrinsic value is not to say that it is something we necessarily pursue as a goal; it is to say that it has its source of value in itself. As Korsgaard puts it, "Intrinsic and instrumental goods should not be treated as correlatives, because they belong to two different distinctions."[64]

It's a pity that Boyle, Gorsuch, and others who argue for the sanctity-of-life principle either are unaware of or simply ignore arguments that refute their simplistic classification scheme. Once we reject their simplistic scheme, we can readily understand why human life is not regarded merely as an instrumental good, but, at the same time, is not a good that must be valued unconditionally in the sense that it must always be preserved, no matter what the condition of a person's life.

Boyle's sanctity-of-life argument fails to come to grips with the central issue of whether life is worth preserving when it is no longer possible to pursue other goods. Many people believe that the value of life is explained by reference to other goods. Obviously, the preservation of life is necessary if we are to fall in love, enjoy friendship, obtain knowledge, work on our personal projects, and so on. However, this does not imply that life is a basic or ultimate good such that we must preserve it when it no longer bears a relation to these other goods, that is, when someone is in such a wretched condition that there is no longer any possibility of working on one's projects, experiencing joy, and engaging in similar worthwhile activities. Instead of addressing this issue squarely, Boyle, Gorsuch, and others circumvent the issue by constructing a result-oriented classification scheme that allows them to place life in the intrinsically valuable category and then claim that it is wrong under any circumstance to end a life. Provided one does not accept their classification scheme—and they provide no good reason to accept this scheme—their argument for the sanctity-of-life principle is wholly unpersuasive.

At one point, Boyle does concede that "people desire not merely to live but to live well, not merely to survive in a vegetative state but to flourish self-consciously in a wide range of goods."[65] However, he persists in maintaining that life is a basic good because even when a human life is in a persistent vegetative state it "can by itself provide a reason for acting."[66] He states, "Family members and health-care workers have chosen to give life-preserving care to [such] persons. . . . *Not everyone* would make such a choice or consider it correct. But the fact that *some* have made it gives evidence that life is a basic good . . ." (emphasis added).[67] There you have it: the ultimate rationale for the sanctity-of-life principle, according to Boyle. Because *some* choose to treat bare biological existence as a good that cannot be destroyed (namely, those who share Boyle's outlook), *all* of us must so regard it.*

Cutting through all the metaphysical jargon, we can see that what Boyle, Gorsuch, and company are really saying is that a terminally ill person who no longer finds existing to be of any value, whose life no longer has any relationship to experiences and goods she finds mean-

*Note that this justification of life as a basic good is also in tension with Boyle's prior characterization of life as an intrinsic good, that is, something desirable for its own sake. Apparently, it is sufficient for something to be a basic good if *some* consider it a reason for their actions. What, then, is not a basic good?

ingful, is morally obliged to refrain from hastening her death merely because *they* think she should. The sanctity-of-life principle serves as a vehicle for imposing the values of some on all; it is a device for appropriating and dictating the lives of others. I do not doubt the sincere commitment of Boyle, Gorsuch, and others who adhere to the sanctity-of-life principle, but, despite their efforts, the principle is intellectually indefensible. The unfortunate result is that the terminally ill suffer for the sake of a platitude.

So far I have shown that the proponents of the sanctity-of-life principle have not been able to develop a positive argument in favor of their view—that it is always wrong to bring about the death of a person intentionally. However, anyone familiar with the literature on this issue will know that positive arguments are not the only weapons used by defenders of the sanctity-of-life principle. Instead, they typically raise an alarm about the consequences of what they characterize as the "quality-of-life" view. They argue that we must adhere to a sanctity-of-life ethic, because that is the only way to maintain the position that human lives are worthwhile. If we adopt a quality-of-life ethic, we will need to make judgments about the quality of individuals' lives and this will place us on the slippery slope to culling human beings based on their capacities or characteristics. Gorsuch decries decisions based on quality of life, claiming this necessarily results in arbitrary judgments about the worth of individuals, and he speculates that the autistic, stroke victims, or those with low IQs will lose the protection of the law under a quality-of-life ethic.[68]

This argument is essentially a scare tactic. All it really does is underscore the tensions and inconsistencies within the sanctity-of-life movement. Any defender of the sanctity of life who also agrees that in some circumstances treatment can be withdrawn or withheld from patients who are no longer competent to make their own decisions (and many of the patients who die as a result of the withdrawal of life-sustaining treatment are not competent to make their own decisions) is implicitly making judgments about an individual's quality of life. To agree with a spouse, a relative, or a friend of the incompetent patient that the burdens of treatment outweigh the benefits to be gained from further treatment is to make a judgment about the quality of the patient's life. For example, if an otherwise robust, healthy patient were admitted to the hospital in an unconscious state for treatment of an acute infection, no physician is going to withhold treatment simply because the spouse says the person should be left to die a "natural" death, and no bioethicist would regard withholding of such treatment to be permissible.* But if that patient were in his late nineties, with multiple serious health issues, including life-threatening cardiovascular disease and terminal cancer, the physician likely would respect the decision of the spouse to withhold treatment. What is the difference between the two cases? The difference obviously lies in a judgment about the quality of life of the two patients. For all those defenders of the sanctity-of-life principle who believe that it is permissible to withhold or withdraw treatment in some circumstances because the burdens of the treatment outweigh the benefits, they are implicitly making judgments based on the quality of life and it is hypocritical of them to deny this fact.

To sum up: we should not use the sanctity-of-life principle to guide our public policy regarding assistance in dying. Applying the principle by its own terms would either require us to prohibit all patients from stopping treatment that would result in their deaths or would force us to resort to distinctions that have no sound, logical basis. Moreover, the defenders of the sanctity-of-life principle are not able to provide any persuasive argument in favor of this principle, and their arguments for this principle are rife with contradictions. Obviously, death is almost always harmful to a person's interests and we should protect people from this harm for that very reason. But when a terminally ill person is in an intolerable condition, then death may not be, on balance, detrimental to that person's interests. We respect both self-determination *and* the obligation to protect people from harm when we allow terminally ill patients to decide for themselves whether they should seek assistance in hastening their death.

---

* I am excluding the situation where the refusal of treatment is based on some religious precept. Even then a physician and hospital would probably not agree without a court order.

THE ARGUMENT AGAINST LEGALIZATION BASED ON DISPARATE IMPACT

We have considered arguments against legalization based on the harm arising out of potential abuses and the sanctity-of-life principle. These are the most prominent ones cited by opponents—and in my opinion the only ones worthy of extended consideration. However, there are other arguments against legalization, including so-called "slippery slope" arguments and those based on the alleged fact that many terminally ill patients who request assistance in dying are depressed. I will address those two arguments briefly at the end of this chapter. I now want to turn to an argument against legalization that one encounters frequently, although not always in a well-developed form. This is the argument that we should not legalize assistance in dying because of the disproportionately adverse, or "disparate," impact the practice allegedly will have on various vulnerable groups.[69]

It may be better to call this a "family" of arguments, because there are several versions of it, with different advocates focusing on different vulnerable groups, and some advocates confusedly blending predictions of future horrible consequences with current social justice concerns. Also, the weight placed on this argument by its various advocates is not uniform, with some including the argument in a list of multiple, apparently similarly persuasive reasons for not legalizing assistance in dying, while others place significant reliance on it, and yet others advance the argument without any clear indication of its importance. Nonetheless, the various versions share a common core: that members of certain vulnerable groups are more likely to be pressured into requesting assisted dying, whether directly by those hostile or indifferent to their interests, or indirectly by social circumstances, such as an inability to pursue other health-care choices.

Probably the most influential statement of this argument is found in the 1994 report of the New York State Task Force on Life and the Law, which counseled against legalization of assistance in dying, in part, because those "most vulnerable to abuse, error, or indifference are the poor

[and] minorities."[70] Certainly the judiciary has been influenced by this argument, as evidenced by the Supreme Court's decisions in *Washington v. Glucksberg* and *Vacco v. Quill*.[71] Both the claims and the perceived strengths of this argument, however, are perhaps best captured in the following succinct summary by bioethicist George Annas:

> The most powerful argument against the legislative expansion of the power of physicians to assist patients in suicide is the danger that this greater latitude will result in abuses that disproportionately affect especially vulnerable populations—the poor, the elderly, women, and minorities. In a country that treats the dying as "freaks," already marginalized members of society could be deprived of their human rights by making them appear somehow less than fully human. This is especially true in the context of cost containment and economic constraints.[72]

Whatever form it takes, the disparate impact argument is fatally flawed. The argument is predicated on social problems that are genuine and disturbing. Members of certain groups are more likely to be treated unsympathetically and are less likely to receive satisfactory healthcare. However, for the disparate impact argument against assistance in dying to be sound, not only must one assume that there is a connection between the social problems just noted and the likelihood of being pushed into assisted dying—a conjecture hitherto totally lacking in empirical support—but one must also make a critical moral assumption. One must believe that it somehow makes a difference for the wisdom of legalizing assistance in dying, whether proportionally more blacks than whites, more women than men, more elderly than young, and so on would likely be pressured into choosing assistance in dying. When one thinks about this proposition, one should be able to see that it is not a morally sound proposition. To the contrary, it is morally repugnant. What matters is whether someone has been coerced into requesting assistance in dying, *not* the race, sex, age, and so on of that person. Unfortunately, disparate impact is reflexively thought by many ethicists and scholars as a sufficient reason for opposing any practice. Disparate impact is as much a dogma as sanctity of life, although its roots are ideological rather than religious.

Note that I am not saying that the risk of coerced or manipulated requests for assistance in dying is not an important factor to consider when determining whether we should legalize it. To the contrary, that is a legitimate concern and I just spent a number of pages evaluating that risk. But the advocates of the disparate impact argument go further and maintain that *how* the risks of legalized assistance in dying are distributed among various groups affects the calculus of benefits and harms from legalization. This is the concern that I question. Although intuitively a society in which all benefits and burdens are distributed proportionally among members of the society's different racial, ethnic, age, and other groups seems more just, we do not insist on such a proportional distribution in all contexts, or even in all contexts analogous to assistance in dying. No one has suggested that competent patients should not be able to request the withdrawal of life-preserving medical treatment because of the disparate impact that the availability of this option might have on certain vulnerable groups, although the risks of this option are similar, albeit not identical, to the risks associated with legalized assistance in dying. Moreover, as discussed in more detail below, it is dubious whether a proportional distribution of the risk of a less-than-voluntary death through assistance in dying would or should satisfy the opponents of legalization.

In surveying the various disparate impact arguments that have been advanced, it is noteworthy that scant attention has been paid to the critical moral assumption just noted, namely, that a disproportionate distribution of risks arising out of legalized assistance in dying argues against legalization. One possible explanation for this omission is that the disparate impact argument functions as a makeweight—an argument thrown into the mix by those uncomfortable with relying exclusively on straightforward arguments that assisted dying is immoral. In other words, the disparate impact argument may well be just another way to advance sanctity-of-life concerns by disguising them in politically fashionable dress. In this regard I note that the majority opinion in *Washington v. Glucksberg* that alluded to the need to protect "vulnerable groups—including the poor, the elderly, and disabled persons—from abuse, neglect, and mistakes" was authored by the late Chief Justice Rehnquist.[73] Say what you will about the former chief justice, but he was never accused of being an overzealous protector of the "vulnerable."

However, surely not everyone who uses the disparate impact argument does so disingenuously. A more plausible alternative explanation for the attraction of this argument is that there is no felt need to justify its key assumptions given some initial skepticism about the wisdom of legalizing assistance in dying. Most ethicists are sufficiently sensitive to the problem of discrimination that, all other things being equal, the fact that a practice or procedure disproportionately burdens certain groups makes the practice or procedure morally suspect. Combine this suspicion with serious misgivings about the wisdom of legalizing assistance in dying and one has what seems to some a ready-made argument against legalization. Thus, opponents who use the disparate impact argument move unhesitatingly from the prediction that legalization will have a disparate impact on one group or another to the claim that this disparate impact, by itself, counsels against legalization.

This move merits scrutiny. There are relevant factual differences between the context of assistance in dying and other social and legal contexts in which the disparate impact argument has been accepted as appropriate. Disparate impact analysis can be a useful means of addressing intolerable inequities when it is applied carefully to achieve an important social goal and there is not only a clear understanding of this goal but a demonstrable causal connection between elimination of the alleged disparate impact and achievement of this goal. Disparate impact analysis has been successfully applied in the context of employment discrimination and housing discrimination. In the employment discrimination context, for example, an employer may not use criteria to screen applicants for a specific job if these criteria are not job related and consistent with business necessity, and use of this selection procedure has a disproportionate impact on one of the groups protected by Title VII of the Civil Rights Act of 1964.[74] In this instance, there is a disparate impact that is clearly defined (members of specified groups are being denied particular, identifiable jobs); use of disparate impact analysis is weighed against the importance and merits of the disputed procedure (the selection procedure is evaluated to determine whether is it necessary for business reasons); and elimination of the disparate impact is causally connected with a goal deemed critical by Congress (equal employment opportunity).

By contrast, in the context of legalization of assistance in dying, there is: (1) a lack of specificity about the disparate impact to be addressed and the goals to be achieved by eliminating or preventing disparate impact; (2) a failure to explain why, on balance, this disparate impact should be a matter of concern; and (3) the absence of a connection between elimination of this impact and critical social goals. All these deficiencies combine to detract from the force of the disparate impact argument.

The oft-cited 1994 report of the New York State Task Force on Life and the Law is illuminating in this regard.* In support of its conclusion that assistance in dying should not be made lawful, the report directs our attention to "employment practices, housing, education, and law enforcement" and reminds us that we have fallen short of eradicating discrimination in these areas. The report warns us that assistance in dying "will be practiced through the prism of social inequality and prejudice that characterizes the delivery of services in all segments of society, including health care." The report then immediately draws the conclusion that legalizing assistance in dying is unwise because those "most vulnerable to abuse, error, or indifference are the poor [and] minorities."[75] Notice that the report provides no clear statement of the goal to be achieved through the prevention of disparate impact, nor any careful consideration of the causal links between elimination of disparate impact and the stated goal. The report also makes no attempt to explain why discrimination in employment or housing, which *denies* members of protected groups the right to pursue desired jobs and housing, is morally equivalent to assistance in dying, which *allows* individuals, including members of protected groups, more freedom of choice in the context of decisions at the end of life. If assistance in dying can be a rational, moral choice for competent

---

* Hardly any opponent of assistance in dying fails to genuflect before this poorly reasoned and skewed report. In reality, it deserves no deference whatsoever. In the assisted dying debate, it's the emperor with no clothes.

persons, then the disparate impact argument (if accepted) erects a barrier to the exercise of this rational, moral choice by anyone, *including* members of disadvantaged groups. What goal does this serve?

The failure to identify the goal being served by use of disparate impact analysis is the most obvious flaw in the disparate impact argument against assistance in dying. Recall that what supposedly concerns the proponents of this argument is that abuses will disproportionately affect vulnerable populations. Could the goal then be equalization of the risk of an improperly hastened death among our nation's various racial, ethnic, and other groups? This goal would be causally connected with the use of disparate impact analysis, just as equalization of employment or housing opportunities is causally connected with use of disparate impact analysis to combat discrimination in those areas. However, acceptance of that goal implies that pressure to request assistance in dying is acceptable as long as, for example, white males are coerced as often as black females. Is this a morally appropriate goal? Should we not be more concerned with ensuring that there are few coerced or manipulated requests relative to the number of persons requesting assistance in dying as opposed to the identity of the persons who are coerced or manipulated?

For example, should we be concerned if it turns out that 75 percent of those pressured into requesting assistance in dying are women, if the absolute number of individuals coerced is low? Would it be worse to have four cases of pressured requests a year if 75 percent of those cases involve women than to have five hundred annual cases of suffering persons being kept alive against their will, assuming their misery is evenly distributed across both sexes? And would it make a difference in our conclusion if somehow we were able to ensure that no more than half of those pressured were women, that is, if two rather than three of the pressured requests came from women? Intuitively, these propositions seem morally dubious, and, in any event, nowhere in the writings of those who use the disparate impact argument can we find any reasoned argument in support of such propositions. Instead, what one finds is a lot of aimless rhetoric lamenting racism, sexism, and the like.

Moreover, if one strips away the rhetoric from these disparate impact arguments, one finds that to the extent they have any force it is because they characterize the option of assistance in dying in starkly negative terms. Assistance in dying is characterized as such an abominable option that the disparate impact analysis becomes irrelevant. An article by Susan M. Wolf applying the disparate impact analysis vividly illustrates this type of argument. After expressing, at some length, her concern with the disparate impact that legalized assistance in dying would supposedly have on women, Wolf states:

> The demand for assisted suicide is a demand for a third party's involvement in purposefully ending a woman's life. That is something women already have in abundance and most people decry. Women are differentially the victims of fatal domestic violence.[76]

Thus, Wolf implies that, for her, assistance in dying is morally equivalent to a husband murdering his wife. To put it mildly, this characterization is tendentious. Moreover, given her view of its morality, Wolf's extended consideration of the sociological impact of assistance in dying becomes extraneous. Wolf should just assert that assistance in dying is as morally objectionable as murder and be done with it.

A simple hypothetical can sometimes uncover an argument's hidden faults. Consider this: Would Wolf's concerns be assuaged if we could assure that men would request assistance in dying at least as often as, if not more than, women? What if we imposed stringent quotas on assisted dying so that no woman would be eligible for assistance in dying unless and until the proportion of men requesting assistance in dying in that year was equivalent to the percentage of women requesting such assistance? Similar restrictions could be imposed for other so-called vulnerable groups: African Americans, the elderly, the disabled, the poor. Quotas would eliminate any disparate impact and, therefore, appear to provide the solution to those concerned about equalization of risk.

It is doubtful, though, that Wolf or any other advocate of the disparate impact argument would accept this proposed solution. Moreover, rejection of this solution would presumably not be due to concern over the surreal bureaucratic regulations necessary to enforce such quotas nor the express limitation on autonomy that results (even with quotas, more persons would have the option of assistance in dying than is now the case). Why then would quotas be rejected? Two reasons come to mind. One reason is that advocates of this version of the disparate impact argument have not heretofore thought through the implications of their argument. Once they recognize that expressly limiting the number of persons from vulnerable groups who could elect assistance in dying would be the most direct way to alleviate their stated concerns about disparate impact, then (I hope) they would also recognize that disparate impact analysis is not useful in this context. In other words, the quotas solution would be rejected because consideration of the quotas solution serves to reveal the flaws in their disparate impact argument.

A second reason the quotas solution might be rejected by some is because the disparate impact of assistance in dying was never their primary concern. People in this group are simply adamant opponents of lawful assistance in dying under *any* circumstances. Although quotas for assisted suicide would resolve their alleged concern, it would not address their real, underlying concern, which is to prevent anyone—black, white, male, female, young, old—from obtaining assistance in dying.[77]

To be fair, some who advance the disparate impact argument do not appear to consider the equalization of the number of pressured choices across groups to be the point of using disparate impact analysis. Instead, the goal—to the extent it is articulated—seems to be raising the standard of medical care so that scarcely anyone from any group would feel subtle pressure to request assistance in dying. For example, Patricia A. King and Leslie E. Wolf state that before we can be confident that patients in vulnerable groups will have an unpressured, uncoerced choice when presented with the option of assisted dying, "changes and modifications . . . are required in the training of healthcare providers and the delivery of healthcare services."[78]

Properly trained healthcare providers and appropriate, adequate, and accessible healthcare for all are certainly worthy and acceptable social goals. It is unclear, however, how, if at all, these goals are connected with maintaining the ban on assistance in dying until we can be assured that assistance in dying will not have a disparate impact on vulnerable populations. It is unlikely, for example, that continuing the ban on assistance in dying will cause the populace to mobilize for well-funded universal healthcare, thus perhaps ensuring that no one will be indirectly pressured into choosing assistance in dying because of his or her inability to afford alternatives. Again, unlike the clear causal connection between the prevention of disparate impact in the contexts of employment or housing opportunities and the acceptable social goals to be achieved, there is a critical missing causal link between prevention of the disparate impact that allegedly will result from assistance in dying and the achievement of acceptable social goals. Continuing to prohibit assistance in dying will not advance us one step toward an equitable distribution of effective medical care. This lack of a connection between a ban on assistance in dying and improved healthcare for the disadvantaged underscores the dubious value of the disparate impact argument in the debate over legalization.

It bears repeating that the evidence from Oregon shows no disparate impact on allegedly vulnerable groups whatsoever. Requests for assistance in dying are evenly distributed among men and women. Almost all the requests have been made by whites (which seems to concern no one). Those seeking assistance are financially secure and, on average, well educated.

But the disparate impact argument is as morally unsound and misguided as it is lacking in empirical support. We should not be dismissive of concerns about prejudice or social injustice. Racism, sexism, and ageism are serious, refractory problems, and nothing I have stated here should be misinterpreted as suggesting otherwise. Nonetheless, continuing to ban assistance in dying will do nothing to solve these problems and there is no morally persuasive reason to support the claim that legalization of assistance in dying should be conditioned on the assurance that members of all groups will incur the same risks of harm following legalization. It is prudent and morally appropriate to implement procedural requirements to ensure that, to the extent possible, all requests for assistance in dying are voluntary; it is unjustifiable to deny everyone—

including a poor, elderly, African American woman—the option of requesting assistance in dying because of a misplaced concern about proportional distribution of risk.

DOPES AND SLOPES

If you can't defeat the argument that competent, terminally ill patients should be allowed, in consultation with a physician, to request assistance in dying—well, just insist that anyone who requests assistance in dying cannot be competent. Or argue that it's not that assistance in dying is bad policy in and of itself, but that it will place us on a slippery slope to some horrible state of affairs, such as assisted dying for healthy teenagers. These are the standard moves one often finds in extended arguments against assistance in dying.

We have already seen how Leon Kass has suggested that terminally ill patients are too dependent on others and susceptible to influence to make a truly informed decision. In other words, they are dopes. Frankly, one might wonder, given the robust procedural safeguards embedded in most proposed statutes for legalization of assistance in dying, how any person without the willpower to run a marathon of bureaucratic hurdles or the mental acuity to craft a credible peace plan for the Middle East could successfully request assistance in dying. In any event, Kass's portrayal of the terminally ill as pliable fools is as insulting as it is inaccurate. Frail bodies do not entail frail minds. Certainly, the evidence from Oregon does not support Kass's supposition. Empirical studies from Oregon indicate that physicians consistently describe those patients who request assistance in dying as having "extremely strong and forceful personalities."[79] Moreover, as we have seen, the data from Oregon show that persons who request assistance in dying are better educated than the average person, and they remain capable of assessing their own situation and acting in accordance with their desires, as indicated by the fact that many choose to forego taking the prescribed medication after they have obtained it.[80]

However, the claim made by opponents is usually not as crude as Kass's contention. The typical argument is not that candidates will have an obvious lack of capacity to make an informed judgment, but that clinical depression will significantly impair the judgment of many of them and that this illness will not be detected by the attending or consulting physicians. Yale Kamisar has urged as one of the primary reasons for not legalizing assistance in dying "the inability of depressed persons to recognize the severity of their own symptoms and the failure of primary physicians to detect major depression in their patients, especially elderly patients."[81] The New York State Task Force on Life and the Law has similarly concluded that "underdiagnosed" depression will result in many incompetent or marginally competent persons opting for assistance in dying.[82]

These concerns are based on dubious conclusions from the available empirical evidence as well as unwarranted assumptions. To begin, there is insufficient evidence to sustain the claim that depression influences persons to request assistance in dying instead of continuing with medical treatment. In addition, opponents who play the depression card make an unwarranted assumption, namely, that "[t]he notion of competence to make treatment decisions . . . also presumes that the patient is not clinically depressed."[83] The fact is that depression in the overwhelming majority of cases neither destroys competency nor inclines a person to an unrealistic assessment of her situation.

Let's first consider the alleged connection between depression and requests for assistance in dying. Empirical studies to date have provided a mixed verdict on the correlation between depression and a desire to hasten death, through assisted dying or otherwise (e.g., by refusing treatment or requesting increasingly heavy sedation), but, on balance, an objective reading of these studies shows no convincing evidence of a causal link between clinical depression and requests for assistance in dying. Some studies of patients outside of Oregon suggest a correlation between depression and patients who *might* request assistance in dying had they the opportunity.[84] On the other hand, some studies have shown no such correlation.[85] More important, in Oregon itself, several studies have shown an absence of any correlation between depression and actual requests for assistance in dying.[86] These studies from Oregon are clearly entitled to the most weight. There is an important distinction between merely thinking about hastening one's death and actually making the request for assistance. As we have already noted, a substantial number of terminally ill patients may think about hastening death; relatively few actually make the request.

Moreover, a couple of important factors must be borne in mind when reviewing these studies. One, diagnosing depression is a tricky matter. There is no lab test or x-ray that will reveal whether someone is depressed. Two, contrary to the claims of those opposed to assistance in dying, it is much more likely that physicians will overdiagnosis rather than underdiagnosis depression in terminally ill patients given the standard criteria for diagnosing depression. The criteria for diagnosing depression are vague and excessively broad, and, moreover, were designed to diagnose depression in physically healthy persons. These criteria encompass attitudes and symptoms that it would not be surprising to find in some terminally ill persons. The criteria for clinical depression set forth in the *Diagnostic and Statistical Manual of Mental Disorders* are as follows:

1. Depressed mood;
2. Markedly diminished interest or pleasure in almost all activities;
3. Significant weight loss/gain;
4. Insomnia/hypersomnia;
5. Psychomotor agitation/retardation;
6. Fatigue;
7. Feelings of worthlessness (guilt);
8. Impaired concentration (indecisiveness);
9. Recurrent thoughts of death or suicide.[87]

Several of the physical symptoms, such as fatigue, insomnia, and weight loss are bound to be present in some of the terminally ill. Even opponents of assistance in dying, such as the New York State Task Force on Life and the Law, recognize this, leading to the advice that "psychological symptoms of depression, such as hopelessness and helplessness, are often more reliable markers than physical symptoms in the assessment and treatment of major depression among individuals with chronic and terminal illness."[88] However, if we then rely only on psychological symptoms, we are sure to be overinclusive: criteria 2 and 9 almost by definition would be satisfied by a person who has decided that in her present condition death is preferable to pursuit of any activities. Furthermore, criterion 1 threatens to be uninformative, functioning more as a restatement of the diagnosis rather than a factor in the diagnosis. Thus, many of those individuals who decide to request assistance in dying are likely to satisfy three of the five psychological criteria used to diagnose clinical depression. They may well be diagnosed as clinically depressed even when they are manifesting attitudes that are not unusual in someone who knows he is approaching death. Someone near death is very likely to have "recurrent thoughts of death."

In addition to the lack of empirical evidence correlating depression with a desire for hastening death, there is no support for the implicit assumption made by opponents that depression impairs a person's judgment to the extent that he is no longer competent to decide on the course of his medical treatment. Although there is some intuitive appeal to the notion that clinical depression necessarily deprives a person of the competence to make medical treatment decisions, "the only evidence that depression is actually associated with decisional incapacity . . . is anecdotal."[89] In contrast, five "empiric studies involving older patients in medical settings have found no significant effect of depression on decision making."[90] One study specifically concluded "that depressed subjects were able to comprehend medical information and to reason as well as the cognitively normal comparison group."[91] Commenting on these studies, one physician has stated they do not "support the veto power which we have allowed depression to have."[92] The President's Commission for the Study of Ethical Problems in Medicine in commenting on this issue concluded that "a diagnosis of a major psychiatric illness only rarely in itself decides the question of the patient's capacity to make a particular treatment decision" and "[t]here is no necessary connection between mental illness and the presence or absence of decisional capacity."[93] Accordingly, the fear that because depression may be "underdetected," numerous incompetent or marginally competent patients will be put to death, appears exaggerated.

Note that I am not arguing that clinical depression can never impair judgment nor that the physicians who are responsible for determining the competency of a person requesting assistance in dying should ignore evidence of depression. Instead, I am merely observing that the possibility that clinical depression may go undetected does not present as great a concern as some opponents of assistance in dying have maintained. Clinical depression and incompetence do not enjoy a one-to-one correspondence.

Nor is there sufficient justification for concluding that depression necessarily distorts a person's judgment, even if it does not render him incompetent. Some have speculated that although a depressed person may retain cognitive capacity, it is possible that depression-induced "helplessness produces an underestimation of one's possible effectiveness in the face of serious illness."[94] In other words, even if clinical depression does not usually produce incompetence, it may skew one's judgment with the result that the depressed will be inclined to request assistance in dying without objective justification. But the notion that the depressed suffer from a distorted perception of reality rests more on intuition than on hard evidence. Not only is there little evidence that the depressed are inappropriately pessimistic in general, but "there is clear evidence that non-depressed people distort reality in a self-serving direction and depressed people tend to see reality accurately."[95] In other words, the depressed do tend to assess situations differently: their assessments are more accurate. They only seem inappropriately pessimistic because the majority of us are inappropriately optimistic.

At the end of the day, the depression argument seems to be just another maneuver to prevent people from making a choice of end-of-life care that some find morally repugnant. Given the ease with which someone can claim a patient's request for assistance is motivated by depression, opponents can label persons who request assistance in dying depressed and, therefore, disqualified from having their request honored. The late Joel Feinberg concisely captured this misuse of a diagnosis of clinical depression as a Catch-22:

> The assumption apparently is that if a depressed person requests to die that *proves* his depression impairs judgment, and his request therefore is insufficiently voluntary to be granted. This argument suggests that only persons who are happy are capable of voluntarily choosing suicide, and of course they are precisely the ones who won't apply. Thus if you are unhappy you *cannot* voluntarily choose suicide, and if you are happy you *will not* commit suicide. The conclusion: no suicide.[96]

That concerns about depressed patients represent another gambit in the efforts to block assisted dying is supported by the stark inconsistency in how the dangers of depression are portrayed in the context of requests for assistance in dying versus the context of requests for refusal of treatment. There is no demand for prohibiting patients from declining life-sustaining treatment, so what procedural safeguards are in place to ensure that refusals of life-sustaining treatment are not made by individuals suffering from clinical depression?

The answer is precious few. As previously indicated, the law recognizes that competent individuals have a right to refuse any sort of treatment, including life-sustaining treatment. Also, for the most part, competency determinations remain in the hands of the attending physician. Consulting specialists are brought in only if the attending physician deems it necessary. Accordingly, many patients are already deciding to terminate treatment, a decision that will probably result in their death, without any assessment of their competency to make this decision other than that carried out by their attending physician. By contrast, the ODWDA includes a requirement that a patient's competency be assessed by the attending physician and a consulting physician.

All the foregoing arguments suggest that the alleged risk of underdiagnosis of clinical depression does not pose as serious a problem as has been suggested by opponents of assistance in dying. There is no correlation between depression and actual requests for assistance in dying; depression does not necessarily render someone incompetent or unduly pessimistic; and among the terminally ill, there is more of a risk of overdiagnosis of depression than underdiagnosis. Finally, there is scant oversight of the psychological condition of patients who refuse lifesaving treatment, yet they run risks comparable to those who would request assistance in dying. Therefore, no one can plausibly maintain that risks from clinical depression among the terminally ill are so great that they warrant a continuation of the ban on assistance in dying.

Another favorite bugbear of opponents of assistance in dying is the dreaded slippery slope, that is, the contention that once the practice of physician assistance in dying is legalized, we will be unable to contain the practice of hastening death to competent, terminally ill patients. Instead, the practice inevitably will be broadened to include anyone who wants to die for any reason, or perhaps we will legalize nonvoluntary euthanasia—the mercy killing of someone not presently capable of making her desires known—or even involuntary euthanasia—the "mercy" killing of someone against her wishes. (I am assuming, for sake of argument, that nonvoluntary euthanasia is a bad thing; depending on one's understanding of nonvoluntary euthanasia and the specific facts at issue, it may not always be morally impermissible.)[97] Exactly which parade of horribles is supposed to follow legalization depends on the opponent's imagination, as does the nature of the slippery slope that is supposed to lead us to this horrible result. In fact, one problem in addressing slippery slope arguments is that opponents often blend, without realizing it, different forms of the slippery slope argument. There are at least four possible slopes: the logical or conceptual slope, in which case horrible practice B—for example, assisted dying for anyone regardless of their condition—is supposedly logically entailed by A—legal assistance in dying for the terminally ill; the precedent slope, in which A is seen as a compelling reason for B, but does not necessarily entail B; the causal slope, in which A leads to B as a result of nonrational changes in attitude; and the full slope, in which both rational and nonrational forces are at work in moving us from A to B. One could easily spend one hundred pages or more discussing the many variants of the slippery slope argument. Here, I will restrict myself to making a few critical observations. These should suffice to show that slippery slope concerns are unfounded or exaggerated and do not provide a sufficient justification for continuing the ban on physician assistance in dying for the terminally ill.

Let us first consider the contention that physician assistance in dying either logically entails or functions as a rational precedent for expansion of assistance in dying to everyone, no matter what the condition. Daniel Callahan has argued that once the "key premises" of the argument for assisted dying are accepted, there will be no logical way to deny assisted dying "to anyone who requests it for whatever reason, terminal illness or not."[98] Along the same lines, the New York State Task Force on Life and the Law (manifesting its penchant for imprecision and fallacious reasoning) maintained that:

> A policy of allowing assisted suicide or euthanasia only when a patient voluntarily requests an assisted death, and a physician also judges that assisted suicide or euthanasia are appropriate to relieve suffering, is inherently unstable. The reasons for allowing these practices when supported by *both* a patient's request and a physician's judgment would lead to allowing the practices when *either* condition is met. . . . In particular cases, and more broadly over time, assisted suicide and euthanasia would be provided based on any serious voluntary request by a competent patient, regardless of his or her medical condition.[99]

Conceivably, this argument could have some force with respect to some rationales for assistance in dying, but not with respect to the one I have offered here (nor with respect to the rationales that most other ethicists have offered). Recall that my defense of the morality of physician assistance in dying and of the removal of legal prohibitions on physician assistance in dying is predicated principally on *all* of the following considerations: (1) prohibiting assistance to a terminally ill patient effectively forces that person to stay alive against his will, thus depriving that person of a fundamental liberty interest; (2) allowing a terminally ill patient to consult openly with his physician about end-of-life options will provide an opportunity for careful discussion and exploration of all alternatives; (3) providing a terminally ill patient with the assurance that he can end his suffering if and when it becomes intolerable actually encourages the person to live longer; (4) and restricting assistance in dying to the terminally ill provides us with some assurance that the person's desire to hasten his death is not an emotional overreaction to some passing problem. These considerations obviously have *no* relevance to, nor do they support, the extension of assistance in dying to those who are healthy enough to kill themselves without difficulty. The notion that this rationale for assistance in dying could be used to support giving a teenager who broke up with her boyfriend a prescription for a lethal dose of a barbiturate is just ludicrous. Assistance in dying for the terminally ill that is based on their special situation and their special needs simply does not logically entail or serve as a precedent for assistance in dying on demand.

What about the argument that assistance in dying for the terminally ill will expand to include other categories of individuals, not as a matter of logic but as a result of a change in attitudes? This type of slippery slope argument maintains that by loosening our inhibitions about hastening the death of the terminally ill, we will end up becoming so callous and insensitive about death that we will become comfortable with assistance in dying on demand or nonvoluntary or involuntary euthanasia. This version of the slippery slope argument is not frivolous. It is also more difficult to refute outright because essentially it is based on speculation about the future. Predictions cannot be proven to be mistaken until the predicted events fail to come to pass—and even then the predictor can argue that she is simply wrong about timing, not the predicted events themselves. Nonetheless, what evidence there is indicates that our footing is secure and we are not in danger of falling into the abyss. Although this version of the slippery slope argument is entitled to some consideration, it is ultimately unpersuasive.

First, in Oregon there has been neither a broadening of the categories of individuals eligible for assistance in dying nor any loosening of the procedures for obtaining assistance in dying. Furthermore, there has been no significant demand to expand assistance in dying beyond the terminally ill. Given that the ODWDA has been in place for a decade, it is remarkable that there has not been one step toward the dreaded slippery slope.

Moreover, as indicated, Oregon remains the only state in which assisted dying is available legally. For those of us who support assisted dying, this is obviously a disappointment. However, the failure to enact comparable legislation in other states does show that the movement for assisted dying is hardly an inexorable, corrosive force that will propel us to a society in which the elderly, the infirm, the emotionally unstable, and the mentally challenged are dispatched with or without their consent. Even if Vermont, California, and a handful of other states enact legislation comparable to the ODWDA in the next few years, assisted dying probably will not be accepted in most states. More important, assisted dying will almost surely be confined to the competent terminally ill. There is hardly any support in this country for legislation that would allow assisted dying on demand or euthanasia in the absence of express consent.

Nor, as long as we remain a democracy, is there likely to be a movement to legalize such practices. The interests of too many people are at stake for us to approve measures whereby anyone can obtain a prescription for a lethal dose of drugs or persons can be euthanized without their consent. Regarding assistance in dying on demand, few people will want such an option available for their loved ones—or themselves—given the well-known dangers of suicidal impulses. Similar interests in self-preservation or the preservation of loved ones make legalization of nonvoluntary or involuntary euthanasia highly unlikely. For example, consider one category of individuals often claimed by opponents of assistance in dying to be at risk, namely, the elderly. Most of the elderly are presumably disinclined to vote themselves out of existence. Furthermore, not only are the elderly going to increase in number as the "baby boom" generation ages, but the elderly are notoriously more likely to vote than younger adults. Regarding other projected victims of the tumble down the slippery slope, such as the cognitively impaired or the disabled, it is important to bear in mind that everyone is potentially a member of these groups. The fact that misfortune can transform anyone into a potential target for nonvoluntary or involuntary euthanasia is likely to act as an impediment to legalizing such practices. All other things being equal, almost all persons want to live. The overwhelming majority of individuals will be disinclined to place themselves in a position where they can be euthanized without their explicit request for such an action.

The fact that we live in a democracy is often overlooked and is also important for assessing the strength of the slippery slope argument in the following respect: Most everyone who employs this argument assumes that the fall down the slope is irreversible. But we have many examples from our history in which we have reversed course when, rightly or wrongly, we have decided that a policy was unwise. In recent decades, divorce laws have been tightened, eligibility for welfare has been narrowed, prohibitions of certain types of pornography have been more vigorously enforced, and access to alcohol by young adults has been restricted. I take no position on any of these issues, but these examples, and many others, illustrate that implementing policies that arguably reflect a change in attitude (more tolerance for divorce, a relaxation of the work ethic, etc.) does not preclude a return to prior ways of thinking. Our citizens do respond to the argument that "things have gone too far." There is little reason to believe there would not be a similar willingness to rethink our policies on assistance in dying if in fact they resulted in practices that many found objectionable.

Finally, one of the flaws in many versions of the causal slippery slope argument is that there is an implicit assumption that the operative attitude is respect for life, and because allowing assistance in dying shows less respect for life, this may eventually lead some to accept the legitimacy of nonvoluntary or involuntary euthanasia. Richard Doerflinger has argued that "Once the taboo against killing has been set aside, it becomes progressively easier to channel one's aggressive instincts into the destruction of life in other contexts."[100] Using much less inflammatory language, Beauchamp and Childress note, in summarizing the causal slippery slope argument:

> If we focus on the modification of *attitudes*, not only on *rules*, shifts in public policy may erode the general attitude of respect for life. Prohibitions are often both instrumentally and symbolically important, and their removal could weaken a set of attitudes, as well as practices and restraints that we cannot replace.[101]

But what is the justification for assuming that the operative attitude of those who favor assistance in dying for the terminally ill is a diminished respect for life? No empirical study with which I am familiar suggests that those who favor assistance in dying have less respect for life or that this is one of the primary motivations behind the efforts to legalize assistance in dying. It seems to me that the primary operative attitudes are respect for self-determination and compassion for the plight of someone who is suffering. Respecting the right of a terminally ill person to make his own decision about the course of his remaining days hardly seems to evince a diminished attitude of respect for life. Thus, a key element of the causal slippery slope argument is missing. The projected progressive erosion of respect for life can hardly take place if it has not even started.

We have been discussing slippery slopes. I would like to end this chapter by employing a different—and, I firmly believe—more appropriate metaphor. Rather than the first step down a slippery slope, I suggest that legalization of assistance in dying for the terminally ill represents another step upward, in our march toward greater respect for the autonomy of the individual. The history of personal liberty in the Western world since the time of the Enlightenment can be described—admittedly, in sweeping, general terms—as a progress toward an ever-increasing enlargement of the sphere of personal autonomy. This progress has been anything but uniform and rectilinear, of course, as Nazi Germany and the Soviet Union remind us. Nonetheless, the advances of the past two hundred years cannot be gainsaid. First among the important advances was religious freedom—the ability to decide for oneself what to believe about the gods (if anything). Many then acquired political freedom—the right to choose what laws to impose on themselves (at least indirectly, via representatives). Enlarging civic and social freedom by, among other things, eliminating slavery, legally emancipating women, and removing formal barriers to upward mobility generally, constituted another set of critical advances. Then, in the last century, we have seen autonomy expanded in the sphere of our personal relations, such as marriage and divorce and reproductive options. Finally, the time has come for an expansion of autonomy in the sphere of our most intimate life and death decisions.

From this overview one can discern that each advance in autonomy made the achievement of each subsequent advance much easier. Each advance has provided a foothold, as it were, for a subsequent step upward. At least in terms of achieving greater personal freedom, we are ascending, not falling into the abyss.

It is interesting, by the way—if I can risk using another spatial metaphor—that respect for autonomy has, in a sense, worked its way inward. Starting with greater freedom in connection with our relationships to deities, we then progressed to greater freedom in our relationships with our rulers, then to our fellow citizens, then to those close to us, and so on. But perhaps this should not be surprising. Toppling dogmatism in religion and authoritarianism in politics are probably necessary conditions for autonomy in other areas. Neither the incentive nor the ability to decide certain matters by and for ourselves would have been present had we always been obliged to follow those who presumed to have the authority to mandate how we should live our lives, whether they be kings, popes, or dictators. In any event, having gained much more control over the course of our lives by being able to decide how we worship, whom we will have as our political leaders, which profession to pursue, whom to marry, and so on, it is now time we obtained some measure of self-determination in how we live out the final chapter of our existence.

Given my extended song of praise for self-determination, one might conclude that I would fully support claims of conscientious objection by healthcare professionals. But remember, autonomy is not a trump. We may have to undertake or refrain from undertaking certain actions because of obligations to others. These obligations may be especially strong when they have been voluntarily assumed and when others are relying on our expertise. It is time to consider the extent to which healthcare professionals should be allowed to decline to provide services to which they object.

NOTES

1. *Oregon Death with Dignity Act*, Oregon Revised Statutes (2003) 127.800–995.

2. Neil M. Gorsuch, *The Future of Assisted Suicide and Euthanasia* (Princeton, NJ: Princeton University Press, 2006).

3. Ibid., pp. 157–66.

4. *Patient Self-Determination Act of 1990, U.S. Code*, vol. 42, sec. 1395cc(f) (2003).

5. *Planned Parenthood v. Casey*, 505 U.S. 833, 851 (1992).

6. *Eisenstadt v. Baird*, 405 U.S. 438, 453 (1972). See also *Loving v. Virginia*, 388 U.S. 1, 12 (1967). Some may regard my citation of these Supreme Court decisions as paradoxical, if not brazen, since the Supreme Court has rejected the claim that there is a constitutional right to assisted suicide. See *Washington v. Glucksberg*, 521 U.S. 702 (1997). But, first, in *Glucksberg* the Court concluded that there was no general constitutional right to receive assistance in dying that could be invoked by anyone. It did not expressly rule on the issue of whether a terminally ill patient may have a right to assistance in some circumstances. My reading of *Glucksberg*, by the way, is similar to Gorsuch's interpretation, and Gorsuch, obviously, is no supporter of assistance in dying. See *The Future of Assisted Suicide*, p. 17. Second, I believe the court failed to follow the implications of its prior decisions in *Planned Parenthood v. Casey* and *Eisenstadt v. Baird.*

7. Joseph Raz, *The Morality of Freedom* (New York: Oxford University Press, 1986), p. 373.

8. Yale Kamisar, "Some Non-religious Views against Proposed 'Mercy-Killing' Legislation," *Minnesota Law Review* 48 (1958): 1011.

9. Richard A. Posner, *Aging and Old Age* (Chicago: University of Chicago Press, 1995), p. 238.

10. See Linda Ganzini et al., "Physicians' Experiences with the Oregon Death with Dignity Act," *New England Journal of Medicine* 342 (2000): 557–63.

11. Oregon Rev. Stat., sec. 127.800(12). See Hospice Care, General Provisions and Definitions, Code of Federal Regulations, title 42, sec. 418.3 (2006) (a person is terminally ill if "the individual has a medical prognosis that his or her life expectancy is six months or less if the illness runs its normal course").

12. Kamisar, "Some Non-religious Views," p. 977.

13. Gorsuch, *The Future of Assisted Suicide*, p. 222.

14. Posner, *Aging and Old Age*, pp. 247–48.

15. Office of the Attorney General, "Dispensing of Controlled Substances to Assist Suicide," *Federal Register* 66 (November 9, 2001): 56607.

16. *Gonzales v. Oregon*, 546 U.S. 243 (2006).

17. By the way, this is not a "how-to" book. I am not going to describe how persons can end their own lives. Perhaps I could sell more books if I did, but I believe it would be inappropriate for me to do so. The interested person with Internet access can inform himself within a few minutes, anyway.

18. These statistics come from a combination of official reports required under the ODWDA and articles that use survey data. The ODWDA reports track the number of prescriptions issued and deaths via prescribed medication, but not patients who consider or request assistance. See Oregon Department of Human Services, *Eighth Annual Report on Oregon's Death with Dignity Act* (2006), pp. 4–5. This report is available at: http://egov.oregon.gov/DHS/ph/pas/docs/year8.pdf (accessed July 4, 2007). For survey information see Linda Ganzini et al., "Physicians' Experiences with the Oregon Death with Dignity Act," *New England Journal of Medicine* 342 (2000): 557–63 (discussing requests from patients and patients who change their minds about following through with their request); Susan W. Tolle et al., "Characteristics and Proportion of Dying Oregonians Who Personally Consider Physician-Assisted Suicide," *Journal of Clinical Ethics* 15 (2004): 111–18 (discussing patients who consider assistance in dying).

19. Gorsuch, *The Future of Assisted Suicide*, p. 1.

20. Lisa Belkin, "Doctor Tells of First Death Using His Suicide Device," *New York Times*, June 6, 1990, p. A1. See also *People v. Kevorkian*, 210 Mich. App. 601, 534 N.W.2d 172 (1995). For Kevorkian's own description of this case, see his book *Prescription: Medicide* (Amherst, NY: Prometheus Books, 1991), pp. 221–31.

21. Ezekiel Emanuel et al., "Euthanasia and Physician-Assisted Suicide: Attitudes and Experiences of Oncology Patients, Oncologists, and the Public," *Lancet* 347 (1996): 1805–10.

22. The requirements set forth in this paragraph are all found in the statute, Oregon Rev. Stat., secs. 127.800–995. See also the *Eighth Annual Report on Oregon's Death with Dignity Act* (see note 18), which provides a description of the ODWDA's procedural requirements.

23. These statistics are from both the *Eighth Annual Report on Oregon's Death with Dignity Act* and the summary information from the statute's ninth year of

operation. Oregon has now stopped issuing detailed reports and provides the relevant data in summary form. See Oregon Department of Human Services, *Summary of Oregon's Death with Dignity Act-2006 and Tables 1 and 2* (2007). The summary and accompanying tables are available at: http://egov.oregon.gov/DHS/ph/pas (accessed July 4, 2007).

24. Summary of *Oregon's Death with Dignity Act-2006 and Tables 1 and 2* (2007).

25. *Eighth Annual Report on Oregon's Death with Dignity Act*, p. 24.

26. Federal Bureau of Investigation, *Crime in the United States 2005*, Expanded Homicide Data, table 9 (2006). This report is available at: http://www.fbi.gov/ucr/05cius/offenses/expanded_information/data/shrtable_09.html (accessed July 14, 2007). See also Callie Marie Rennison and Sarah Welchans, *Intimate Partner Violence*, a special report prepared by the Department of Justice, Bureau of Justice Statistics, May 2000. The report is available at: http://www.ojp.usdoj.gov/bjs/pub/pdf/ipv.pdf (accessed July 14, 2007).

27. Did I just pull this ratio out of my hat? Yes and no. As already indicated, there is no precise risk-benefit analysis one can make here. Even if we make the implausible assumption that each terminally ill person is fungible (all terminally ill persons have the same values, preferences, expectations, responsibilities, etc.), and even if we simplify matters by ignoring the data suggesting that making assistance in dying lawfully encourages some people to live longer, we are effectively balancing the benefit of being able to forego living in intolerable conditions for a certain limited period of time (say, one month) against the harm of being manipulated into dying somewhat earlier than one otherwise would have (say, one month). There is no satisfactory metric that I am aware of that would allow us to make a precise comparison between this set of benefits and this set of harms. Nonetheless, it seems to me that a roughly 10 percent ratio of undesirable side effects to desirable results would make us question the wisdom of most actions, especially if those side effects are serious, as they are here.

28. Don Phillips, "Federal Speed Limit, Set in 1974, Repealed," *Washington Post*, November 29, 1995, pp. A1, A16.

29. US Department of Health and Human Services, *The Third National Incidence Study of Child Abuse and Neglect* (Washington, DC: GPO, 1996), pp. 3–12. The *Fourth National Incidence Study* is under way, with the results expected to be released in December 2008, after the writing of this book has been completed. See https://www.nis4.org/schedule.asp (accessed June 19, 2007). If possible, any

significant differences between the third and the fourth studies will be noted in the published book. Currently available statistics indicate the number of annual deaths from child abuse remains roughly the same. See Child Welfare Information Gateway, *Child Abuse and Neglect Fatalities: Statistics and Interventions* (Washington, DC: GPO, 2006), p. 2 (estimated fatalities: 1,490).

30. US Department of Health and Human Services, *A Nation's Shame: Fatal Child Abuse and Neglect in the United States* (Washington, DC: GPO, 1995), pp. 9, 18–19.

31. Ibid., p. 17.

32. *Third National Incidence Study*, p. 5–4.

33. Ibid.

34. See, for example, Hugh LaFollette, "Licensing Parents," *Philosophy and Public Affairs* 9 (1980): 182–97.

35. *Eisenstadt v. Baird*, 405 U.S. 438, 453 (1972).

36. Claire C. Obale, *Patient Care Decision-Making: A Legal Guide for Providers* (St. Paul, MN: West Group, 1997), p. 8–3.

37. For a study discussing the lack of appropriate documentation in cases involving withdrawal of treatment see Karin T. Kirchoff et al., "Documentation of Withdrawal of Life Support in Adult Patients in the Intensive Care Unit," *Journal of Critical Care* 13 (2004): 328–34.

38. Leon R. Kass, "'I Will Give No Deadly Drug': Why Doctors Must Not Kill," in *The Case against Assisted Suicide: For the Right to End-of-Life Care*, ed. Kathleen Foley and Herbert Hendin (Baltimore: Johns Hopkins University Press, 2004), p. 24.

39. "The taboo against physician-assisted death is crucial," according to Kass. "Why Doctors Must Not Kill," p. 29.

40. Pope John Paul II, *The Gospel of Life* (New York: Random House, 1995), p. 102.

41. *Washington v. Glucksberg*, 521 U.S. 702 (1997) and *Vacco v. Quill*, 521 U.S. 793 (1997).

42. *Vacco*, p. 800.

43. Ibid., p. 801.

44. Ibid., pp. 801–802. The Court quoted from the testimony of Leon Kass. House Committee on the Judiciary, *Assisted Suicide in the United States*, 104th Cong., 2nd sess., 1996, 367.

45. *Vacco*, p. 802.

46. Gorsuch, *The Future of Assisted Suicide*, p. 52.

47. Ibid., p. 66.

48. Ibid.

49. The clearest statement of this view is found in Jorge L. A. Garcia, "Better Off Dead?" *APA Newsletter on Philosophy and Medicine* 92 (Spring 1993): 85–88. See also Edmund D. Pellegrino, "Doctors Must Not Kill," *Journal of Clinical Ethics* 3 (1992): 96. Pellegrino claims that assistance in dying cannot be justified out of respect for the patient's self-determination because the dead patient will no longer be able to make choices about anything. Again, this is just sophistry. Choosing to put an end to one's misery is a choice, even if one can no longer make further decisions after that choice. Pursuing Pellegrino's reasoning to its ultimate conclusion would imply that we should not honor wills or other testamentary instructions because, after all, what difference do decisions about the disposition of one's property make to the person once she's dead? That Pellegrino, a learned, distinguished bioethicist, makes such a gross error in reasoning indicates the strong emotional opposition that many have to assistance in dying; in some cases, this emotion blinds opponents to logic.

50. Jacob Axelrad, *Patrick Henry: The Voice of Freedom* (New York: Random House, 1947), pp. 110–11.

51. Joseph Boyle, "Sanctity of Life and Suicide: Tensions and Development within Common Morality," in *Suicide and Euthanasia*, ed. Baruch A. Brody (Dordrecht, Netherlands: Kluwer, 1989), pp. 221–50.

52. See Gorsuch, *The Future of Assisted Suicide*, pp. 157–66.

53. Boyle, "Sanctity of Life and Suicide," p. 232.

54. See Tom L. Beauchamp and James F. Childress, *Principles of Biomedical Ethics*, 5th ed. (New York: Oxford University Press, 2001), p. 129.

55. Samson's death is described in Judg. 16:25–30.

56. St. Augustine, *City of God*, trans. Henry Bettenson, 1.21.

57. Augustine declined to invoke DDE in Samson's case, but he excused Samson on the ground that God gave him a secret command. This excuse, by the way, shows the poverty of morality based on alleged divine commands. Why cannot every person who commits suicide or hastens his death maintain that God has given him a secret command?

58. David Orentlicher, "The Supreme Court and Physician-Assisted Suicide—Rejecting Assisted Suicide but Embracing Euthanasia," *New England Journal of Medicine* 337 (1997): 1237.

59. Helga Kuhse provides a lengthy analysis of DDE in *The Sanctity-of-Life Doctrine in Medicine: A Critique* (Oxford: Clarendon Press, 1987), pp. 83–165. For

more abridged treatments, see also Timothy E. Quill, Rebecca Dresser, and Dan W. Brock, "The Rule of Double Effect—A Critique of Its Role in End-of-Life Decision Making," *New England Journal of Medicine* 337 (1997): 1768–71; James Rachels, *The End of Life* (New York: Oxford University Press, 1986), pp. 15–17, 92–96, 104–105.

60. Boyle, "Sanctity of Life and Suicide," p. 236.

61. Ibid., p. 237.

62. Gorsuch, *The Future of Assisted Suicide*, p. 158.

63. Christine M. Korsgaard, "Two Distinctions in Goodness," *Philosophical Review* 92 (1983): 169–95.

64. Ibid., p. 170.

65. Boyle, "Sanctity of Life and Suicide," p. 237.

66. Ibid., p. 238.

67. Ibid., pp. 238–39.

68. Gorsuch, *The Future of Assisted Suicide*, pp. 159–62, 172–76.

69. Some who have advanced this argument include: Gorsuch, *The Future of Assisted Suicide*, pp. 125–27; Marc S. Spindelman, "Legislating Privilege," *Journal of Law, Medicine and Ethics* 30 (2001): 24–33; Patricia A. King and Leslie E. Wolf, "Empowering and Protecting Patients: Lessons for Physician-Assisted Suicide from the African-American Experience," *Minnesota Law Review* 82 (1998): 1015–43; Susan M. Wolf, "Physician-Assisted Suicide, Abortion and Treatment Refusal," in *Physician-Assisted Suicide*, ed. Robert F. Weir (Bloomington: Indiana University Press, 1997), pp. 167–201; New York State Task Force on Life and the Law, *When Death Is Sought: Assisted and Euthanasia in the Medical Context* (New York: 1994), pp. 125–26; and George J. Annas, "Physician-Assisted Suicide—Michigan's Temporary Solution," *New England Journal of Medicine* 328 (1993): 1573–76.

70. New York State Task Force on Life and the Law, *When Death Is Sought*, p. 125.

71. *Washington v. Glucksberg*, 731; *Vacco*, pp. 808–809.

72. Annas, "Physician-Assisted Suicide—Michigan's Temporary Solution," p. 1575.

73. *Washington v. Glucksberg*, 731.

74. See *Griggs v. Duke Power Co.*, 401 U.S. 424 (1971) and *Albemarle Paper Co. v. Moody*, 422 U.S. 405 (1975). See also *Civil Rights Act of 1991, U.S. Code*, vol. 42, sec. 2000e–2(k)(1)(A)(i) (2003), which codifies certain aspects of a disparate impact claim, including the business necessity defense.

75. New York State Task Force on Life and the Law, *When Death Is Sought*, p. 125.

76. Wolf, "Physician-Assisted Suicide, Abortion and Treatment Refusal," p. 177.

77. In a spirited commentary on one of my published articles on the disparate impact argument, Marc S. Spindelman contends that proponents of the disparate impact argument are not really concerned about proportional numbers, but rather are concerned about the reasons women as opposed to men, blacks as opposed to whites, etc., would request assistance in dying. In other words, the disparate impact argument should be interpreted as being concerned with "differential influence" rather than a "differential distribution" of coerced deaths. See Spindelman, "Legislating Privilege," pp. 24–29. (This was a response to my article "Should We Impose Quotas? Evaluating the 'Disparate Impact' Argument against Legalization of Assisted Suicide," *Journal of Law, Medicine and Ethics* 30 [2001]: 6–16.) There are a couple of problems with Spindelman's claim, not the least of them being that several of the disparate impact arguments explicitly invoke numbers, not reasons. See New York State Task Force on Life and the Law, *When Death Is Sought*, pp. 125–26. In addition, how could we possibly determine whether there had been "differential influence" without looking at the numbers, especially since Spindelman and others argue that we cannot always accept a patient's stated reasons as the true reasons for his or her request? For an expansion of these points and other arguments see my "The Need to Specify the Difference 'Difference' Makes," *Journal of Law, Medicine and Ethics* 30 (2001): 34–37.

78. King and Wolf, "Empowering and Protecting Patients," p. 1043.

79. Tolle, "Characteristics and Proportion of Dying Oregonians Who Personally Consider Physician-Assisted Suicide," p. 116.

80. Ibid. See also *Eighth Annual Report on Oregon's Death with Dignity Act.*

81. Yale Kamisar, "Are Laws against Assisted Suicide Unconstitutional?" *Hastings Center Report* 23 (May–June 1993): 38.

82. New York State Task Force on Life and the Law, *When Death Is Sought*, p. 127.

83. Ibid., p. 126.

84. See, for example, William Breitbart et al., "Depression, Hopelessness, and Desire for Hastened Death in Terminally Ill Patients with Cancer," *Journal of the American Medical Association* 284 (2000): 2907–11.

85. See, for example, Steven M. Albert et al., "Wish to Die in End-Stage ALS," *Neurology* 65 (2005): 68–74.

86. Linda Ganzini et al., "Interest in Physician-Assisted Suicide among Oregon Cancer Patients," *Journal of Clinical Ethics* 17 (2006): 27–38; Linda Ganzini et al., "Experiences of Oregon Nurses and Social Workers with Hospice Patients Who Requested Assistance with Suicide," *New England Journal of Medicine* 347 (2002): 582–88; Linda Ganzini et al., "Physicians' Experiences with the Oregon Death with Dignity Act," *New England Journal of Medicine* 342 (2000): 557–63.

87. American Psychiatric Association, *Diagnostic and Statistical Manual of Mental Disorders*, 4th ed. rev. (Washington, DC: APA, 1994), pp. 320–27.

88. New York Task Force on Life and the Law, *When Death Is Sought*, p. 16.

89. Melinda A. Lee and Linda Ganzini, "Depression in the Elderly: Effect on Patient Attitudes toward Life-Sustaining Therapy," *Journal of the American Geriatrics Society* 40 (1992): 983.

90. Ibid. The actual quote from the article refers to "four empiric studies," but the article itself represents the fifth such study.

91. Ibid. The reference is to Barbara Stanley et al., "The Functional Competence of the Elderly at Risk," *Gerontologist* 28 (1988): 53–58.

92. Erich H. Loewy, "Of Depression, Anecdote, and Prejudice: A Confession," *Journal of the American Geriatrics Society* 40 (1992): 1069.

93. President's Commission for the Study of Ethical Problems in Medicine and Biomedical and Behavioral Research, *Deciding to Forego Life-Sustaining Treatment* (Washington, DC: GPO, 1983), p. 123n7.

94. Mark D. Sullivan and Stuart J. Younger, "Depression, Competence and the Right to Refuse Lifesaving Medical Treatment," *American Journal of Psychiatry* 151 (1994): 974.

95. Martin E. P. Seligman, *Learned Depression* (New York: Random House, 1991), p. 111. See also Lauren B. Alloy and Lyn Y. Abramson, "Depressive Realism: Four Theoretical Perspectives," in *Cognitive Processes in Depression*, ed. Lauren B. Alloy (New York: Guilford, 1988), pp. 223–65.

96. Joel Feinberg, *Harm to Self* (New York: Oxford University Press, 1986), pp. 359–60.

97. Because a discussion of euthanasia is tangential to the principal arguments of this chapter, I have not provided a precise definition of "euthanasia." Most persons have a sufficient understanding of euthanasia—roughly, bringing about the death of another person in order to end that person's suffering—to realize why nonvoluntary euthanasia (under many circumstances) and involuntary euthanasia might be considered practices to be avoided. For an attempt at a rigorous defini-

tion of euthanasia, see Tom L. Beauchamp and Arnold I. Davidson, "The Definition of Euthanasia," in *Moral Problems in Medicine*, 2nd ed., ed. Samuel Gorowitz et al. (Englewood Cliffs, NJ: Prentice Hall, 1983), pp. 446–58. I would argue that nonvoluntary euthanasia, mercy killing without the contemporaneous explicit consent of the person whose death is brought about, is morally permissible in some circumstances. The reader will recall my example of Major Hayward from chapter 2. Nathaniel Poe did not have an opportunity to consult with Hayward prior to shooting him, but the possibility that Hayward would have objected to Poe's actions is infinitesimally small. Turning to less unusual circumstances, the deaths of thousands of incompetent patients are brought about each year through withdrawal or withholding of treatment. Many have, not implausibly, characterized such actions as "passive" euthanasia, yet this practice is considered both morally permissible and legally acceptable, provided certain procedures are followed. "Active" euthanasia, that is, deliberately bringing about the death of a person other than through the withdrawal or withholding of medical treatment, for example, by injecting a drug that will cause a quick, painless death, is more controversial, but it seems to me that in some circumstances it would be morally permissible, even when the patient is not presently capable of providing consent. Consider, for example, a terminally ill patient in obvious pain, but without sufficient cognitive capacity to request assistance in dying, who nonetheless, when he had sufficient cognitive capacity, executed an advance directive that gave his consent to euthanasia under precisely those circumstances in which he now finds himself. Nonetheless, I also believe that effective regulation of active euthanasia for even this subset of incompetent patients would be exceedingly difficult. Active euthanasia would be very susceptible to abuse. Again, we must bear in mind the distinction between what is permissible morally and what should be permissible legally. Anyway, the situations that those who employ the causal slippery slope argument have in mind when they refer to the horrors of nonvoluntary euthanasia are usually the killing of the insane, mentally retarded, and severely disabled incompetents. These situations, of course, bear no resemblance to situations of terminally ill patients eligible for assistance under the ODWDA.

98. Daniel Callahan, *The Troubled Dream of Life: Living with Mortality* (New York: Simon & Schuster, 1993), p. 107. Note that Callahan was addressing voluntary euthanasia, since that was the principal option being discussed at the time he wrote this passage, but I believe it is fair to read this passage as also expressing concerns about the type of assistance in dying allowed under the ODWDA.

99. New York State Task Force on Life and the Law, *When Death Is Sought*, pp. 100–101.

100. Richard Doerflinger, "Assisted Suicide: Pro-Choice or Anti-Life?" *Hastings Center Report* 19 (January–February 1989): S19.

101. Beauchamp and Childress, *Principles of Biomedical Ethics*, p. 146. I note two things: First, Beauchamp and Childress neither endorse nor reject the causal slippery slope argument. After their summary, they observe that the strength of the argument is difficult to assess, given its speculative character. Second, their uncritical acceptance of the view that the attitude at issue is respect for life is a rare (perhaps unique) error in an otherwise excellent and tightly argued book.

4.

My Choices—Your Conscience?

After a long flight, you have finally made it out of the terminal and are eager to get home. It is only a short taxicab ride away. You climb into a cab, give the driver directions, and wait for the cab to begin moving. It doesn't. Instead, the driver asks you what is in the bag you are carrying. It is a bottle of wine. The driver informs you that, for reasons of conscience, he cannot give you a ride home. Disgusted, you climb out of the cab. It is the same story with the next few drivers, all of whom happen to be conscientious Muslims who strongly object to alcohol. Finally, you are successful in finding a driver who does not object to the items you choose to take home with you. The next day you call the taxicab commission to complain.

Now let's consider a different scenario. You are in the hospital with your siblings. Your father, who is effectively brain dead, is being kept alive by a ventilator. Pursuant to his express written directive, and in agreement with your brothers and sisters, you decide to have the ventilator removed. In fact, you and the rest of your family have all gathered at his bedside to be there when he dies. There is just one problem. No one on the staff wants to help you. All of the nurses refuse to remove the ventilator on grounds of conscience. Frustrated and distraught, you and your family complain to the hospital administrator.

Is there any difference between these two cases? In both situations, you have service providers refusing to provide their normal services on grounds of conscience. The actions of the nurses in the second scenario are arguably more objectionable, given the stress and emotional turmoil caused by their actions. However, it turns out that the principal difference between these two cases is that in the first situation (which is based on actual events at the Minneapolis–St. Paul International Airport), your complaint will likely have an effect on the taxicab drivers. They will be informed that to maintain their license, they cannot turn away passengers merely because they are carrying alcohol.[1] In the second situation, your complaint may well fall on deaf ears. Many states and the federal government have now provided healthcare workers—and some healthcare institutions—with "conscientious objector" status in a number of different situations, allowing them to refuse to participate in procedures such as withdrawal of life-sustaining treatment, sterilization, assisted reproduction, and abortion.[2] Proponents of conscientious objector exemptions have tried to get healthcare workers excused from participating in other specified procedures, such as organ transplants and blood transfusions. The latest, but surely not the last, claim to conscientious objector status comes from pharmacists who refuse to dispense emergency contraception, also known as "Plan B."

But in addition to exemptions from specified procedures, there are general exemptions provided by several states that allow healthcare workers *and facilities* to refuse to participate in *any* procedure to which they object on grounds of "conscience." Both Illinois and Mississippi, for example, have very broad conscience clauses that allow *any* healthcare provider to refuse to participate in *any* procedure and *any* healthcare institution to refuse to provide *any* service. Regarding institutions, the relevant provision of the Mississippi code states that: "A health care institution has the right not to participate, and no health care institution shall be required to participate in a health care service that violates its conscience."[3]

Here is a question for you: How do you determine the "conscience" of an institution?

Time was that one could rely on any given nurse, pharmacist, or hospital to provide needed services. Now conscientious objector exemptions are proliferating more rapidly than stem cells under the Bush administration.

How did we arrive at this situation? Until the 1970s, American law, with rare, irrelevant exceptions, recognized conscientious objector (hereafter "CO") status only for individuals legally compelled to join certain groups with the consequence that they were obliged to perform actions to which they had strong religious objections. The Quaker drafted into the military or the Jehovah's Witnesses required to attend school are familiar examples. The former objects to participating in war; the latter to certain school exercises, such as saluting the flag. Either by statute or case law interpreting the Constitution, special exemptions were recognized for such individuals.

Then came *Roe v. Wade*. The Supreme Court's decision legalizing abortion resulted in a wave of statutes, both federal and state, that allowed physicians and other healthcare providers as well as healthcare institutions to decline to participate in abortion. Other statutes covering other services soon followed, so we now have a complicated situation in which, depending on the state in which they are located, various categories of healthcare providers (or institutions) may refuse to provide different types of services, with different degrees of protection for the healthcare worker. In some states, for instance, healthcare workers have the right to sue for damages for an adverse employment action resulting from just such a refusal of care.

I believe that most healthcare workers in most instances should *not* have a right to refuse to provide healthcare services. No, the police should not be called in to move the nurse's hand so she can disconnect the ventilator. No one should be forced to do a job under threat of a prison sentence, for prudential reasons if nothing else. But neither should the nurse be able to refuse to exercise her responsibilities and keep her job. Conscience should not be cost free, especially when one's conscience interferes with another's choice. Bear in mind that no one is forcing a pharmacist to take emergency contraception herself; neither is anyone forcing a nurse to author an advance directive requesting termination of his own life support. What some healthcare workers want to do is to refuse to provide services requested by others on the ground that they are somehow responsible for the healthcare decisions of others. The notion that healthcare workers are responsible for the decisions of others is simply untenable, and healthcare workers should not be permitted to obstruct the healthcare decisions of others. If my pharmacist truly believes he is responsible for my healthcare decisions, will he pay for my medication?

Regarding the right of institutions to claim CO status, I would characterize this purported right as a myth, but it doesn't deserve even that much credit. It's just nonsense. Institutions do not have a conscience and we do not have a moral obligation to respect their "autonomy." Moreover, if they are funded by the public—and almost all hospitals affiliated with religious groups receive public funding—they should be required to address the needs of all the public, not just those with whom they are in philosophical agreement. As a pragmatic, political compromise, there may have been some basis for allowing religious hospitals to refuse to provide abortions and some other services for a period of time after *Roe v. Wade*. That decision was not entirely expected, and we cannot expect hospitals to change their policies overnight. But *Roe v. Wade* was decided over thirty years ago. Hospitals should not be allowed to continue to feed at the public trough while picking and choosing among the services to be provided.

Of course, all I have done so far is state my views. I need to support my views with argument. Let us now turn to that task.

NORMS AND FACTS RELATING TO CONSCIENTIOUS OBJECTION

At bottom, the dispute over CO status—at least as regards individuals—can be analyzed as a balancing of the autonomy rights of the healthcare worker against the autonomy rights and needs of the healthcare recipient and the public good. In some ways, this is a replay of our discussion of assistance in dying, but there is one very fundamental difference: the person who is requesting assistance in dying is trying to exercise some control over *her* life. Those healthcare workers who seek CO status are not only trying to maintain their personal integrity, but are also interfering with the decisions *of others*. This distinction is critical and has often been overlooked, which is one reason our policies on CO status are so misguided.

Following my recommended procedure, let us first set forth some accepted norms and background facts that may provide us with considered judgments relevant to the issue of CO status.

First, as we established in the prior chapter, there is a prima facie obligation to respect someone's choices. The presumption that, all other things being equal, we should defer to a person's choices is especially strong in the context of what we have referred to as critical life choices, that is, choices that provide shape and direction to one's life. Examples of such choices include decisions on marriage, childbearing, career, and the like.

Decisions based on deeply held religious convictions share some of the characteristics of critical life choices. For those committed to a particular religious belief, that belief is typically extremely important to them. It is no exaggeration to say that for the deeply religious, their lives are built around their faith. Admittedly, decisions about discrete issues, such as whether to abstain from certain foods or to wear certain articles of clothing, may seem trivial to those of us outside a particular faith. But we recognize the importance of these issues to those within the faith.

Centuries of bitter experience have taught us that requiring someone to renounce her faith, either by formally abjuring her faith or by undertaking an action inconsistent with her faith, is almost always an enterprise that is counterproductive, whatever our motivation for coercing the person to renounce her faith. Often, the person will resist vigorously, perhaps to the point of violence. Furthermore, with respect to certain issues, such coercion seems pointless, because the actions are primarily self-regarding. As Thomas Jefferson once observed, "it does me no injury for my neighbor to say there are twenty gods or no god. It neither picks my pocket nor breaks my leg."[4] It makes little or no difference to me whether a Jew wears a yarmulke or refuses to eat pork, a Muslim kneels in prayer or refuses to drink alcohol, a Pentecostal speaks in tongues or refuses to read a Harry Potter book, and so on. Accordingly, for actions that are self-regarding, there is a presumption we should respect a person's decision to undertake or refrain from undertaking a certain action because of that person's religious belief. In the United States, certain protections for the "free exercise of religion" are built into our legal system, but even absent such protections it would be both morally objectionable and pointless, for example, to force a Jehovah's Witness to recite the Pledge of Allegiance. We gain nothing from the forced participation in a ceremony that a person inwardly rejects—other than perhaps the satisfaction of seeing someone bend to our will.

Of course, not all situations involving a person's conscientious objection to certain actions are so easily resolved. In some cases, a person's actions or inactions will affect others, and as we also established in the previous chapter, to the extent someone's conduct affects others, those effects must be taken into account in deciding whether we should respect that person's autonomy. The extent to which conscientious objection should be respected is an issue that our national and state governments have grappled with since the beginning of our nation, as there has always been a substantial body of individuals, principally Quakers, who have refused to take up arms on the grounds of conscience.[5]

What factors might weigh in favor of granting CO status to someone who objects to taking certain actions when his refusal adversely affects others? First, if the conflict between the person's conscience and the obligation to take the action is not a conflict that the person herself has generated, then that factor weighs in favor of recognizing CO status. For example, in the case of someone objecting to military service, that factor suggests that an exemption might be appropriate since the person did not choose to become a member of the military. Instead, the government, through the draft, has compelled that person to become a member of the military.

Another factor that weighs in favor of recognizing CO status is the person's willingness to accept some burden, in particular some form of substitute service, as a result of refusing to provide required services. This factor is important because it may balance out the harm that the person is causing by failing to provide services. In addition, it is a way of testing the sincerity of the person's conscientious objection. If the person truly refuses to provide services on grounds of conscience, she should be willing to endure some reasonable burden to compensate for her failure to provide the services in question.

Furthermore, if the exemption from required service can be carefully circumscribed so that the harmful effects of the refusal are limited to the extent possible, then this also argues in favor of granting CO status. Again using the military example, a CO may be excused from service in the military, but he is not allowed to obstruct others. A CO is not permitted to block a military convoy.

Finally, there is a pragmatic consideration. Would forcing someone to engage in a particular action produce a desired result or is it likely to be counterproductive? Again, let's consider the military situation. There is little point in forcing someone to bear a rifle if he is not going to use it, and the costs associated with requiring religious objectors to become soldiers or face prison probably outweigh any possible benefit. Moreover, the military service obligation can still be discharged effectively by other individuals, so creating an exemption does not unduly interfere with the objective of the law making military service compulsory.

Based on the foregoing factors, we can make the judgment that where: (1) the conflict between conscience and service obligation is created by the government; (2) the CO is willing to accept some burdens as the price of conscience, in particular some form of substitute service; (3) the exemption from service can be narrowly tailored so the CO's refusal does not obstruct the decisions of others; and (4) there are pragmatic reasons for granting CO status, then there is a reasonable basis for honoring a conscientious objection to providing certain required services. All the above factors have been considered and accepted by the government as bearing on the wisdom of allowing persons who conscientiously object to all forms of participation in war an exemption from military service. The person who refuses to serve in the military for religious reasons did not create the conflict—he or she was drafted; to obtain CO status, a CO must accept alternative service; the CO is not allowed to obstruct the work of those who are performing military service; and the cost to the government of doing without the military services of the CO is probably outweighed by the cost of forcing the CO to participate.

CONSCIENTIOUS OBJECTION BY HEALTHCARE WORKERS

By contrast, none of these factors, with the possible exception of the last one, is applicable to healthcare workers who conscientiously refuse to provide certain services.

To begin, no one is forced to become a physician, pharmacist, or nurse. Unlike a conscripted Quaker, these individuals have chosen to place themselves in a situation where they may be required to provide services to which they object. Unless and until the government legally requires certain individuals to become physicians, pharmacists, or nurses, these professionals are in a totally different situation than a person who is drafted into the military or compelled to attend school and salute the flag.

Some physicians might argue that there is an implicit understanding that they have the discretion to provide or refuse to provide services, except in emergency situations, so any infringement of that discretion is at least analogous to being required to serve in the military. It is true that many physicians have traditionally exercised some control over the patients for whom they accept responsibility and the services that they provide. A physician in private practice may limit the number of patients that she accepts and a dermatologist may decline to carry out a colonoscopy. Whether the discretion that physicians have often enjoyed is sufficient to ground a claim for conscientious objection is questionable, but we will note it now as one possible reason for distinguishing physicians from other healthcare professionals. In any event, pharmacists, nurses, and other healthcare workers have not traditionally limited their "practices." Once the physician and patient have decided on a course of treatment, the pharmacist and the nurse are supposed to assist in this treatment plan, not create obstacles to its fulfillment.* The situations of the pharmacist and nurse bear no resemblance to the conscripted pacifist.

---

*Pharmacists and nurses are, of course, not automatons. If, based on their specialized knowledge and skills, they believe that a proposed course of treatment is too risky to the patient or not efficacious, they are expected to present their concerns. But that is different than refusing to provide a service based on conscientious objection.

Not only have healthcare workers volunteered to provide services to the public, but the most important categories of healthcare workers also have successfully lobbied the State to grant them monopolistic privileges over these services. If you are not a physician you cannot prescribe medication, and if you are not a pharmacist, you cannot dispense medication. Whether nurses have a true monopoly is questionable, but access to many nursing services is tightly regulated and controlled. Effectively, one cannot obtain healthcare in this country without the cooperation of physicians, pharmacists, and nurses. "Conscientious" physicians, pharmacists, and nurses want to exercise control over our access to healthcare *and* deny us healthcare when they see fit. Allowing physicians, pharmacists, and nurses to have exclusive control over most critical aspects of our healthcare while simultaneously retaining the right to deny these services from time to time based on their personal beliefs is analogous to allowing critical military command positions to be staffed by Quakers and then leaving it to their discretion to decide, without penalty, whether their units will be deployed for combat.

In evaluating the first factor to be considered in evaluating a claim for conscientious objection—that is, whether the person asserting the claim has had any role in creating the conflict between her conscience and the requirement to provide services—it is apparent that this factor does not favor granting an exemption to most healthcare workers. Evaluation of the second factor—whether the CO is willing to accept alternative service or incur some sort of penalty—similarly cuts against granting an exemption.

It is striking that virtually all those healthcare workers and their representatives who have argued for the right to refuse to provide services have also argued that this refusal should not be accompanied by a substitute service requirement or any sort of penalty. The nurse who refuses to assist in the withdrawal of lifesaving treatment maintains he does not have to work an extra shift or assume the responsibilities of the nurses who are willing to provide the requested services. The pharmacist who refuses to dispense emergency contraception argues that she should not be required to find another pharmacist who can assist the patient she turns away. Indeed, Karen Brauer, president of Pharmacists for Life, maintains not only that pharmacists who refuse to dispense certain medication should enjoy complete job protection, but also that they should not be required to refer a patient to another pharmacist or even to transfer the prescription.[6] Unlike the military CO, healthcare workers want their act of conscientious objection to be cost free. They want others to respect their conscience, but they are unwilling to incur the slightest cost to protect their integrity themselves. Instead, they want all the costs to be borne by the patient.

These costs are not just hypothetical. In a notorious case in 2002, a University of Wisconsin student requested a refill of her prescription for an oral contraceptive. The only pharmacist on duty at the Kmart pharmacy was Neil Noesen, a Roman Catholic. Noesen refused to fill the prescription and also refused to transfer the prescription to another pharmacy. The prescription was not filled for over two days.[7] This incident does not stand alone. There were approximately two hundred similar incidents involving pharmacists during a six-month period in 2004.[8]

One excuse given for failing to assist the patient through a referral to another physician, pharmacist, or nurse who might be willing to help the patient is that such a referral makes the healthcare worker complicit in an immoral action. As Brauer has stated, requiring pharmacists to transfer a prescription is "like saying 'I don't kill people myself but let me tell you about the guy down the street who does.'"[9] Somewhat more eloquently, Edmund Pellegrino has argued that it is "illicit" for a physician who conscientiously objects to some procedure to engage in "active cooperation in finding a physician who will provide the morally objectionable service."[10] In other words, the physician can just wash her hands of the patient. This insistence by healthcare workers who believe they are entitled to CO status that they should not lift a finger to help the patient underscores why the third factor we listed above—whether the harm created by the refusal to serve can be minimized—appears to weigh against granting healthcare workers CO status, at least on the terms they desire. Pursuant to a misunderstanding of personal responsibility, the healthcare CO does not want to do *anything* that even indirectly assists the patient who is being denied services.

This insistence by some healthcare workers that they not only may but must leave patients to fend for themselves rests on a fundamental misunderstanding about personal responsibility and the limited exemption traditionally granted the CO. A CO is not allowed to obstruct others. We excuse objecting students from saluting the flag, and, similarly, we allow pacifists to perform alternative service; however, we do not allow objecting students to remove all flags from the school nor do we allow pacifists to block military convoys. The attempt by healthcare workers to analogize their situation to traditional conscientious objectors does not withstand scrutiny. If we required all female pharmacists to take emergency contraception after sexual activity, then they could justifiably protest that such a requirement violates their personal integrity. A decision about whether to take such a drug is clearly a highly personal one, and should be left to the discretion of the individual. But healthcare workers are not complaining about government requirements that affect their bodily integrity or their right to self-determination; instead they insist they have a right to decide *for others* what treatment should be available. This exaggerated sense of personal responsibility for the decisions of others would lead to absurd consequences if workers in other areas could refuse to provide services in the same way that some healthcare workers claim they can refuse to provide services. A librarian could refuse to stock books by noted atheists because she would be "complicit" in the immoral conduct of those who choose to read such books. A flight attendant could refuse to serve kosher meals on the plane because this constitutes "active cooperation" with anti-Christian beliefs. The vegan social worker could refuse to distribute food stamps to families who eat meat and also refuse to refer these families to another caseworker because such a referral would be "illicit." And, yes, Muslim taxicab drivers could refuse to transport passengers who are carrying alcohol. Such refusals to provide services do not constitute conscientious objection as traditionally understood. Rather, they bespeak a more primitive taboo mentality in which a person somehow becomes morally tainted by associating with others who engage in actions the person would not engage in himself.

This misunderstanding of the scope of personal responsibility is underscored by a Supreme Court decision that addressed the claim of a person who objected on religious grounds to the issuance, and the government's use, of a social security number. Distinguishing the case of the military CO, the Supreme Court ruled that the Court had never "interpreted the First Amendment to require the Government itself to behave in ways that the individual believes will further his or her spiritual development or that of his or her family." Respecting an individual's conscience "simply cannot be understood to require the Government to conduct its own internal affairs in ways that comport with the religious beliefs of particular citizens."[11] Similarly, no individual should be forced to undergo a medical procedure to which she objects, but a healthcare worker does not have a right to refuse to assist in some procedure or dispense a medication that a patient needs; patients should not be required to limit their healthcare to comport with the religious beliefs of healthcare providers.

Admittedly, once again, physicians may be in a different category than other healthcare professionals. Ideally, the physician and the patient will work together in planning a course of treatment. Moreover, although the degree to which patients can choose their own physician has eroded considerably in recent decades, they still exercise considerably more control over this choice than over the nurse or pharmacist who the hospital or drugstore assigns to help them. Thus, we may have some reason to accept a physician's claim that she bears some responsibility for a patient's decision. Whether that is sufficient to allow a physician to claim CO status is an issue we will postpone until we have considered all the factors bearing on this question.

In fact, we are at the stage where it is appropriate to evaluate the significance of pragmatic considerations in deciding whether to grant CO status to healthcare professionals. When discussing CO status in the military context, we pointed out that one practical reason for not forcing a pacifist to bear arms is that, presumably, he would not fire his weapon anyway. Moreover, having someone opposed to combat as your comrade in a foxhole is a situation that is less than ideal. For similar reasons, you probably do not want a physician adamantly opposed to abortion performing your abortion, nor do you want a physician adamantly opposed to in vitro fertilization overseeing your efforts at assisted reproduction. Simplifying greatly, the force of these pragmatic considerations varies on a sliding scale that runs from physicians to nurses to pharmacists and then to other healthcare workers. The physician is the person most directly responsible for your healthcare and is usually the person either carrying out or directing the performance of the procedure in question. Nurses do carry out many procedures, but typically at the direction of a physician. The pharmacist's specialized knowledge is useful in determining proper dosage of medication, but she does not have the authority to decide on a patient's course of treatment, nor does she have direct contact with a patient's body. Nurses' aides and other healthcare workers have the least level of responsibility for a patient's healthcare.

In addition, of all these healthcare workers, the physician is the one most likely not to be an employee. Although the independence of physicians has declined markedly in the last few decades (indeed, a majority of physicians are now employees), a substantial number still have their own practices, either on their own or in an arrangement with other physicians.[12] On the other hand, although one still encounters the occasional pharmacist who is independent, this is increasingly uncommon. Nurses and other healthcare workers rarely have their own practices. The fact that many physicians have independent practices whereas most other healthcare workers are employees is relevant because, at least in this country, there is a presumption that a person has a right to operate his business as he sees fit (within certain parameters), whereas employers usually have the right to direct the work of their employees. (Some of the qualifications on this right will be examined in the next section.) Allowing independent physicians, but not employee physicians, to be able to decide what services they will provide poses the risk of confusing the public as well as creating a two-tier system among physicians.

Let us now evaluate all the factors we have considered. It seems apparent that pharmacists, nurses, and most other healthcare workers are not in the same situation as the military CO. They are not compelled to take up their profession, they are not willing to provide alternative service or incur some sort of penalty, their refusals interfere with the decisions of others, and practical considerations do not weigh in favor of granting an exemption. For physicians, the balance is a bit different. Although no one is compelled to become a physician, physicians have traditionally enjoyed a substantial amount of discretion with respect to the services they provide. In addition, they share a measure of responsibility for a patient's healthcare decisions. Furthermore, practical considerations weigh in favor of not requiring a physician to perform a procedure to which he is opposed. Finally, as noted, a patient who does not regard a physician's refusal to provide certain services acceptable generally has more opportunity to find another physician than she does to find an alternate nurse or pharmacist—we typically get the pharmacist or nurse that someone else (a store, a hospital, etc.) has assigned to us. In sum, there is a basis for recognizing a limited right of refusal by physicians.

It is imperative, however, that a physician notify a patient as soon as possible of any limitations on services she provides. Ideally, this should be done via written disclosure at the outset of a relationship so the patient can decide then and there whether to enter into a professional relationship with the physician. Prompt disclosure imposes virtually no burden on the physician and would likely have little effect on her practice. Not every patient will want to reserve the right to request contraception, an abortion, and so on. If a physician has not previously disclosed a limitation on services to the patient, but then refuses to provide such services, she should both be required to assist the patient with a prompt referral to other physicians and should be disciplined for her prior failure to disclose. Conscientious objection that materially interferes with a patient's access to services should not be tolerated, especially when this interference is due to the negligence of a physician.

In terms of crafting a law that would acknowledge a right of conscientious refusal for physicians, a general exemption is preferable to specified exemptions for certain procedures. Specific exemptions are subject to change and create more uncertainty, as well as being an invitation to acrimonious political debate. As long as the law mandates that a patient be informed which procedures the physician conscientiously objects to performing, a general exemption is not likely to result in a shortage of needed services. Physicians who "conscientiously" object to providing rectal exams are not likely to remain competitive.

In any event, once the patient and physician have decided on a course of action, they should be able to rely on the cooperation of other healthcare workers. The last thing we need is to complicate our healthcare system even further by allowing pharmacists, nurses, and others to obstruct a person's healthcare decisions based on their sectarian beliefs. Especially given the increasing religious diversity of our country, we do not want to force patients seeking care to navigate among Catholic nurses, Muslim pharmacists, Wiccan nurses' aides, and Hindu dieticians, and so on, all of whom claim the right to refuse to provide care. We should not require patients to sacrifice their healthcare to assuage the conscience of those who have volunteered to provide this healthcare.

CONSCIENTIOUS OBJECTION BY EMPLOYEES IN OTHER OCCUPATIONS

We do need to determine, however, whether our denial of the right of conscientious refusal to most healthcare workers is consistent with the way in which other employees are treated. It is—to the extent the law in this area is predictable.

The statutes that provide a specific right to conscientious refusal are limited almost exclusively to healthcare workers. However, federal civil rights laws and most state fair employment statutes prohibit discrimination based on religion, just as they prohibit discrimination based on race, gender, national origin, and various other protected classifications. In applying this prohibition of religious discrimination, federal law and the laws of most states have been interpreted to allow employees to claim some sort of accommodation for their religious beliefs. Here, I will limit my discussion to federal law, in part for space considerations and in part because many states tend to follow federal law in interpreting their fair employment statutes.

Title VII of the Civil Rights Act of 1964 is the relevant federal statute.[13] Under Title VII, an employer has a duty to accommodate an employee's religious beliefs, but that duty does not require an employer to incur more than de minimis costs. In the Supreme Court decision on this issue, *Trans World Airlines, Inc. v. Hardison*, the Court concluded that the employer did not have to respect Larry Hardison's religiously motivated request not to work on Saturdays, because finding a substitute for Hardison would have required the employer to deprive other employees of their shift preferences (no one volunteered to take Hardison's shift) or hire another employee to work Saturdays. The Court reasoned that Title VII does not compel an employer to "deny the shift and job preference of some employees . . . in order to accommodate or prefer the religious needs of others" or to incur higher wage costs.[14] Applying this logic, almost every court that has confronted this issue has determined that an employer is not required to retain an employee who categorically refuses to perform essential job functions on the basis of religion. For example, a trucking company can fire a driver who refuses to make long-haul overnight runs with a female partner,[15] the state police do not have to retain an officer who refuses to carry out his assigned duties at a casino,[16] and an employee assistance program does not have to retain a counselor who refuses to advise homosexual clients.[17]

This reasoning has been applied in the healthcare context as well. The US Court of Appeals for the Third Circuit upheld the termination of a nurse who refused to assist in treating a patient who required an emergency caesarian section, which would have terminated the pregnancy.[18] Significantly, the outcome of this case might have been different if the employer and her lawyer had invoked the New Jersey Conscience Statute, which provides that no person shall be required to assist in the performance of an abortion. However, they neglected to make this claim.

Under Title VII, employers are required to determine whether a no-cost accommodation is possible, but that is the extent of their obligation. For example, in the counselor case, the employer offered to transfer the employee to a vacant noncounselor position, but because the position paid less, the employee refused the offer. In light of the employee's decision to decline this reasonable accommodation, the employer was entitled to discharge her.

From the foregoing, we can see that most employers are not required to respect an employee's conscientious refusal to perform assigned duties to the extent it imposes some cost on them. There is no reason that healthcare workers should receive preferential treatment; they do not constitute a privileged class. They should receive no greater protection than that given to other employees under Title VII. If a nurse refuses to participate in the withdrawal of life-sustaining treatment, she can be offered a transfer to some vacant position for which she is qualified that would not require her to come in contact with patients. The hospital should not be required to let her keep her current position, pick and choose among her assigned duties, and interfere with the healthcare decisions of others.

CONSCIENTIOUS OBJECTION BY INSTITUTIONS

Those who favor a broad right of refusal for healthcare workers often emphasize the importance of moral integrity. They argue that protecting the moral integrity of the individual benefits all of us because we want the members of our community to recognize and act upon their moral obligations and "acting conscientiously is the most fundamental of all moral obligations."[19]

There is some force to this point. As I have argued, however, even if we concede there is a need to protect and foster moral integrity, this is an insufficient reason for granting healthcare workers a broad right of refusal. Among other things, by choosing to become physicians, pharmacists, and nurses, they are responsible for generating the conflict between their conscience and their obligations toward patients, and their personal integrity is not directly implicated by the healthcare decisions of others.

But whatever force the argument from integrity has for recognizing a right of conscientious refusal by individual healthcare providers, it has no relevance to the issue of whether institutions such as hospitals, HMOs, and medical clinics should be allowed a right of "conscientious" refusal.

Institutions have no conscience, nor does it serve any public purpose to grant them CO status. No matter how lofty and exalted the principles in a corporation's mission statement, the corporation itself has no moral commitment to anything. It certainly has no religion, and thus its non-existent religious beliefs cannot be offended when its employees eat meat, salute the flag, or dispense contraception.

Notwithstanding the absurdity of granting CO status to institutions, both the federal government and many state governments have done just that. In 1973 Congress passed the first federal law (called the Church Amendments, after the senator who sponsored them, not after the organizations that benefited from them) granting "entities" receiving certain federal funds the right to refuse to make their facilities available for an abortion or sterilization if such a procedure is "prohibited by the entity on the basis of religious beliefs or moral convictions."[20] The legislation was motivated in part by the Supreme Court's decision in *Roe v. Wade* and in part by a lawsuit that attempted to require a hospital controlled by the Catholic Church to carry out a sterilization. Approximately thirty states have since adopted similar provisions applicable to healthcare institutions. Most of these statutes provide a right of conscientious refusal for certain specified procedures, but a few states have very broad statutes that permit institutions to decline to provide any service to which they object.

Other than pragmatic, political reasons, there is no justification for any of these exemptions—narrow or broad. As indicated, they serve no purpose remotely connected to moral integrity. Moreover, most healthcare institutions receive public support of one sort or another, via assistance with initial financing, tax breaks, payments for Medicare and Medicaid patients, and so on. To hold one hand out for public support and to use the other hand to turn away disfavored patients is intolerable. If a healthcare institution chooses to receive public support, then it should be required to provide the public with all legal services it is capable of providing.

"Conscience" exemptions for healthcare institutions are especially troubling given the large number of institutions controlled by religious organizations. Institutions affiliated with churches and other religious organizations constitute a majority of the nonprofit healthcare providers in the United States. Many of these institutions are affiliated with the Catholic Church, which condemns contraception, abortion, sterilization, and other widely accepted healthcare services and procedures. There is nothing wrong with religious bodies operating and controlling healthcare institutions; there is something wrong with such institutions receiving public support if the services they provide are going to be limited by religious dogma. One of the rare nineteenth-century cases to address an Establishment Clause challenge to public support for a religiously affiliated institution considered the constitutionality of funding for Providence Hospital in the District of Columbia.[21] The Supreme Court ruled the funding was perfectly permissible because the hospital provided secular services and the fact that its management was under the control of a Catholic religious order did not imply its healthcare services would be subject to or limited by ecclesiastical authority. This was the correct decision—based on the facts presented to the Court. But if hospitals are going to limit their services based on religious doctrine, then they should not receive public funding. Hospitals should deliver healthcare, not dogma.

There was a practical justification for the Church Amendments. We cannot expect hospitals to change their policies overnight. If the Church Amendments had provided religious hospitals and healthcare institutions with a grace period of five to ten years in which they could decide to get out of the healthcare business or stop receiving public funding, that legislation would have been acceptable. But allowing such institutions the option of continuing to receive taxpayer support while limiting their services as they see fit has no justification, and this approach will just lead to more demands for exemptions and further confrontations over healthcare in the future as additional procedures become accepted by many while remaining unacceptable to some. Will Catholic hospitals that receive public support be allowed to refuse to provide patients with therapies derived from embryonic stem cells?

To sum up: all other things being equal, we should not force an individual to undergo some procedure or take some medication that she objects to on grounds of conscience, but her right to decide for herself does not allow her to decide for others, especially when she has freely chosen a profession that obliges her to assist others. Allowing publicly supported healthcare institutions to refuse to provide services on the basis of a conscience they do not possess is a policy so incoherent, so lacking in logic, that even Kafka would have rejected it as excessively absurd.

So far, we have been evaluating policies where the potential harms and benefits are fairly apparent. But how do we evaluate risk when the potential harms and benefits may not be fully known? That will be a critical question in the next chapter, which discusses regulation of genetically engineered plants and produce.

NOTES

1. See Kari Lydersen, "Some Muslim Cabbies Refuse Fares Carrying Alcohol," *Washington Post*, October 26, 2006, p. A2; John Reinan, "Taxi Proposal Gets Sharp Response," *Minneapolis-St. Paul Star Tribune*, February 27, 2007, p. 1.

2. A good summary of state laws on healthcare workers' rights to conscientious refusal in the context of reproductive procedures (abortion, contraception, sterilization) can be found in Guttmacher Institute, *State Policies in Brief: Refusing to Provide Health Services* (July 1, 2007), http://www.guttmacher.org/statecenter/spibs/spib_RPHS.pdf (accessed July 25, 2007). For an overview of both federal and state conscience clauses applicable to healthcare workers and institutions, see Martha S. Swartz, "'Conscience Clauses' or 'Unconscionable Clauses': Personal Beliefs versus Professional Responsibilities," *Yale Journal of Health Policy, Law, and Ethics* 6 (2006): 269–350, and Mathew White, "Conscience Clauses for Pharmacists: The Struggle to Balance Conscience Rights with the Rights of Patients and Institutions," *Wisconsin Law Review* (2005): 1611–48.

3. *Health Care Rights of Conscience Act of 2004*, Mississippi Code (2007), sec. 41-107-7. See also Illinois, *Health Care Right of Conscience Act*, Illinois Compiled Statutes (2006) sec. 745–70.

4. Adrienne Koch and William Peden, eds., *The Life and Selected Writings of Jefferson* (New York: Random House, 1944), p. 275.

5. Because compulsory military service was first required by individual states, it was the states that first granted exemptions for conscientious objectors (COs). These exemptions were always conditioned on payment of fees or some alternative service requirement. See Lillian Schlissel, *Conscience in America: A Documentary History of Conscientious Objection in America, 1757–1967* (New York: E. P. Dutton, 1968), pp. 56–57. The federal government confronted the issue of conscientious objection to military service during the Civil War, when the draft was first instituted. Ibid., pp. 88–89. The government decided to exempt religious COs. The federal government has continued to grant CO status whenever it has instituted compulsory military service. Note that although the statute exempting a CO from military service states that the objection must be grounded on religious beliefs, to preserve the constitutionality of the statute, the Supreme Court has ruled that an atheist could be a CO if he had an unconditional opposition to war maintained with the strength of traditional religious convictions. *Welsh v. United States*, 398 U.S. 333 (1970).

6. Brauer's position is summarized in Rob Stein, "Pharmacists' Rights at Front of New Debate," *Washington Post*, March 28, 2005, p. A1.

7. This summary is borrowed from White, "Conscience Clauses for Pharmacists," pp. 1611–12.

8. Editorial, "Moralists at the Pharmacy," *New York Times*, April 3, 2005, p. A12.

9. Ibid.

10. Edmund D. Pellegrino, "The Physician's Conscience, Conscience Clauses, and Religious Belief: A Catholic Perspective," *Fordham Urban Law Journal* 30 (2002): 240.

11. *Bowen v. Roy*, 476 U.S. 693, 699 (1986).

12. Many physicians who have their own practices may technically be employees of the professional corporation that they have created. Whether or not a shareholder in a professional corporation is also an employee is sometimes a difficult question to resolve. See *Clackamas Gastroenterology Associates, P.C. v. Wells*, 538 U.S. 440 (2003). However, for my purposes, I consider such physicians to be in control of their practice and not, in a functional sense, employees.

13. *Civil Rights Act of 1964, U.S. Code*, vol. 42, secs. 2000e–2000e-15 (2003).

14. *Trans World Airlines v. Hardison*, 432 U.S. 63, 80–81 (1977).

15. *Weber v. Roadway Express*, 199 F.3d 270 (5th Cir. 2000).

16. *Endres v. Indiana State Police*, 349 F.3d 922 (7th Cir. 2003).

17. *Bruff v. North Miss. Health Services, Inc.*, 244 F.3d 495 (5th Cir. 2001). As previously indicated (see note 3), Mississippi now has a very broad conscientious objection provision for healthcare workers. The result of this case may well have been different under this statute, which was enacted in 2004.

18. *Shelton v. University of Med. and Dentistry of New Jersey*, 223 F.3d 220 (3d Cir. 2000).

19. Farr A. Curlin, "Caution: Conscience Is the Limb on Which Medical Ethics Sits," *American Journal of Bioethics* 7 (2007): 31.

20. *U.S. Code*, vol. 42, sec. 300a-7 (2003).

21. *Bradfield v. Roberts*, 175 U.S. 291 (1899).

5.

Food Fears versus Food Facts—The Need for Appropriate Risk Assessment

Suppose someone excitedly told you the following: "Our milk is being contaminated with bacteria!" "They are marketing meat from genetically modified chicken!" "Stores are selling fruit that has been sprayed with a potent neurotoxin!" "Millions are drinking a fungus-infected beverage that alters brain function!" Given the concerns that have been raised about "processed," "synthetic," or "genetically modified" foods, you might well believe these breathless admonitions are the latest horror stories about some Frankenfood being churned out Avaricious Agrotech, Inc. The next time you are shopping at your local organic food store, stocking up on some wholesome products, you console yourself with the thought that at least you are wise enough to avoid harmful, artificial food.

If that is your reaction to these seemingly dire warnings, you may be fooling yourself. You will find at your friendly organic food store products that match all the descriptions set forth above. Bacteria have been used to make cheese for quite a long time; the chickens we have today have been genetically modified through selective breeding since they were first domesticated; organic farmers can and do use pesticides that are toxic in sufficient concentration; and using yeast to produce beer was a technique first discovered in ancient Egypt. If "organic" food is supposed to designate food that one could find in the wild in a state unmodified by human intervention, then you will not be able to locate much of that at an organic food store or anywhere else for that matter. Perhaps some berries and mushrooms could be so categorized, as well as some game and fish, but that is about all. Ever since humans broke ground with their hoes and began cultivating crops and domesticating animals, we have altered flora and fauna, sometimes deliberately and sometimes accidentally. The notion that organic food is "natural" because it is derived from plants and animals as one might find them in an unaltered state is just a myth.

A FACT-RICH DIET

This chapter will be structured a little differently than other chapters. Before we even arrive at the policy issue to be addressed in this chapter—how to regulate genetically engineered plants and foods—a number of misperceptions about organic and genetically engineered foods require corrections. Despite the strong convictions many have about assistance in dying, conscientious objection by healthcare workers, stem cell research, and so on, in researching this book I discovered that far more inaccuracies and misleading statements surround controversies over our food than any other dispute. There is something about this controversy that causes advocates on all sides of the debate to indulge in hyperbole and myth-making. (Was it something they ate?) As just one example of how even well-respected organizations can go off course, the Union of Concerned Scientists (which is very skeptical about the benefits of genetically engineered foods and promotes organic farming) has this statement on its Web site:

> No one tries harder than organic producers to give American consumers choice in the market place.[1]

Is this a scientifically supported statement? Did the Union of Concerned Scientists conduct a study of the working habits of organic producers and those who engage in other types of farming? What is the definition of "trying hard"? What is the definition of "choice"? Biologists at Monsanto may think that they are working very "hard" to try to give consumers more choice by developing new varieties of plants. Not to put too fine a point on it, this statement is just propaganda, and there is a surfeit of similar empirically unsupported statements being made in the debates over food safety. So bear with me: before I arrive at the policy discussion of plant regulation, I first need to do a substantial amount of brush clearing.

Let's pick up where we stopped. I stated that the notion that organic food is derived from unaltered plants and animals is just a myth. So too is the notion that organic food is healthier for you. No scientific study has ever made that conclusion. Likewise, the notion that organic farming is more respectful of our environment is more rhetoric than reality. Of course, "respectful" is not itself a scientific term, so one needs to explicate that term before separating false claims from accurate statements. If by "respecting" the environment, one means leaving it in as pristine a state as possible, then we passed the point of no return millennia ago. Except for designated wilderness areas, the forest primeval has disappeared. Agriculture of any sort—organic or otherwise—alters the environment to suit human needs. If by "respecting" the environment, one means limiting one's alteration of the environment to the extent possible, then it is at best dubious whether organic farming respects the environment more than other methods of farming. Organic farming generates yields lower than conventional farming, which means it needs more land to produce an equivalent amount of crops; organic farming's reliance on mechanical tillage as its primary means of controlling weeds leads to soil erosion; organic farming's rejection of bioengineering results in pollution that could otherwise be avoided.[2] Organic farming does interact with the environment in ways that differ from conventional farming or farming that uses biotechnology, but it has negative effects on the environment just as these alternative methods do. They are just different negative effects.

The whole concept of organic farming as somehow being in a completely different category than other types of farming because it is more in harmony with nature rests on a misunderstanding of chemistry. Organic farmers reject—for the most part—so-called synthetic substances by which they mean anything created in a lab. (I use the qualifier "for the most part" because organic farmers are permitted to use some synthetic substances, including synthetic pesticides, under certain conditions.)[3] But what if the lab produces a substance that is identical in its chemical composition to a substance used by organic farmers? Scientifically, there is no distinction between the substances. They are composed of identical molecules. Nonetheless, for those in the organic movement the lab-created substance is considered "synthetic."

In fact, the organic farm movement had its origins in the eccentric views of a British agronomist, Sir Albert Howard, and an Austrian mystic, Rudolf Steiner, both of whom rejected use of fertilizers created in labs, even though they were chemically indistinguishable from the active ingredients in so-called natural fertilizers, principally manure.[4] Howard claimed that "artificial" fertilizers would inevitably produce "artificial" men and women.[5] The failure of this prophecy's fulfillment does not seem to have dissuaded Howard's followers from accepting the wisdom of his advice.

The scientifically unsound belief that the process by which a product is made necessarily results in fundamental, occult changes to the product, even when it is molecularly indistinguishable from a product created by a different process, explains why the Department of Agriculture's regulations for organic farming provide that organic farming is defined by the process it utilizes, *not* the chemical composition of its products. The lack of any scientific basis for distinguishing natural from synthetic products also explains why the department's regulations define "natural" in terms of what is not "synthetic" and "synthetic" in terms of what is not "natural." Here are the circular definitions from the Code of Federal Regulations:

> Nonsynthetic (natural). A substance that is derived from mineral, plant or animal matter and does not undergo a *synthetic* process. . . .

> Synthetic. A substance that is formulated or manufactured by a chemical process . . . except that such term shall not apply to *naturally* occurring biological processes. (Emphasis added)[6]

Something is natural if it is not synthetic, and something is synthetic if it is not natural. Got it?

The foregoing summary of some of the misperceptions, myths, and pseudoscience surrounding organic farming is not meant to be an indictment of organic farming. To the contrary, just because some of the claimed benefits of organic foods are wildly exaggerated does not imply that organic food is bad for you or that organic farming methods are substantially more harmful than other methods. Far from it. Although there may be no substantial benefit to organic farming, there is no substantial harm resulting from organic farming either. Some have suggested that organic farmers' reliance on manure creates a greater risk of contamination by *E. coli* bacteria, but there is no reliable evidence confirming this supposition.[7] Organic food does get contaminated, but so do foods grown by other methods. The risk of contamination varies from grower to grower, from food processor to food processor. Nor do organic farming techniques otherwise present any significant concern. As indicated, crop yields from organic farming do tend to be lower, so if all farmers turned organic we could suffer from a food shortage. But all farmers are not going to go organic. Although organic farming has grown rapidly in the last few decades, organic farms still constitute a very small fraction of North American agriculture (less than 1 percent of cultivated acreage).[8] Finally, in defense of organic farming, many devotees of organic food swear it tastes much better—a claim that, regrettably, I cannot confirm from my own personal experience, but, as the saying goes, there's no disputing tastes.

The problem is not with organic farming itself, but with the cult of the organic and the mind-set it produces among a vocal minority of its supporters. In particular, some of its supporters are adamantly opposed to genetically engineered plants and produce, which has caused unnecessary concern about the safety of such plants and produce and has resulted in demands for unnecessary and burdensome regulation, if not a complete ban. Furthermore, this unreasonable aversion to genetically engineered food is based on false claims that genetically engineered food is fundamentally different from and less safe than food currently produced on organic farms or conventional farms.

Let's begin with the use of the term "genetically modified," which is the term most opponents of genetically engineered food use to describe products that utilize bioengineering techniques. This term is misleading because most of the products we eat already, including many of the products we find in organic food stores, have been genetically modified. For millennia, new genetically different plants have been created through crossbreeding species. The end result of this process can produce a plant that looks nothing like, and has characteristics completely different from, its predecessors. Corn is derived from a weed, teosinte, that originally had but a few hard seeds, and modern-day corn bears scant resemblance to this weed.[9]

Crossbreeding is an inexact technique that sometimes has resulted in plants with desirable traits and sometimes in plants with undesirable traits. Even when the plant produced appears to have traits the cultivator deems appropriate and suitable, it nonetheless may have properties, such as allergens, that prove dangerous to consumers. Kiwi, for example, is a fruit developed through "natural" crossbreeding in New Zealand during the twentieth century. After it was marketed, a number of consumers had allergic reactions.

Kiwi, corn, and a host of other products are derived from genetically modified plants, yet they all can be found in almost any supermarket, including organic food stores. No special labels are placed on them to indicate they are genetically modified. The scaremongers confine their use of the term "genetically modified" to plants that have been genetically engineered, that is, plants whose genetic composition has been changed through the introduction or elimination of specific genes through recombinant DNA technology. The implication of singling out genetically engineered food for the scare label of "genetically modified" is that genetically engineered plants are produced by a radically new and risky process. But the National Research Council, an affiliate of the National Academy of Sciences, has clarified that genetic engineering is merely "one type of genetic modification."[10]

Furthermore, there is no reason to think that genetically engineered food is inherently less safe than food produced through traditional genetic modification of plants. Genetic engineering differs from crossbreeding principally in two respects, namely, the precision and range of the genetic modification resulting from the two techniques. Let's discuss precision first, which in almost every conceivable circumstance constitutes an advantage. Genetic engineering is the targeted manipulation of genetic material. The genes that are transferred are few in number and carefully selected, unlike the haphazard approach of crossbreeding, which can introduce hundreds of unwanted genes into the hybrid. In traditional crossbreeding, these unwanted genes have to be eliminated through "backcrossing," that is, breeding the hybrid with the original parent crop over several generations.

This is an appropriate place to note, by the way, that traditional crossbreeding and genetic engineering are not the only ways to produce novel plants. New plants can also be produced through mutagenesis, in which plants are intentionally bombarded with radiation, scrambling their genes and producing mutants. Hundreds of useful and commercially viable crops have been produced by mutagenesis, such as much of the red grapefruit from Texas. Mutagenesis also has produced thousands of totally useless, if not dangerous, mutations. Mutagenesis is an unpredictable approach to plant modification; some have likened it to firing a shotgun at a genome. Mutagenesis actually has much greater potential for producing unintended compositional changes in a plant than either crossbreeding or genetic engineering.[11] Nonetheless, it has scarcely raised a whimper of concern among those opposed to genetically engineered plants. And yes, you can buy mutated produce from organic food stores, with no warning label attached.

Let's now turn to the other principal difference between crossbreeding and genetic engineering, namely, the range of potential modifications. Unlike traditional genetic modification, which is limited largely to crossbreeding between related species,[12] genetic engineering can transfer genes from any other organism into another organism. Accordingly, genetic engineering can produce novel plants, and food products derived from such plants, that could not be achieved by other methods of genetic modification. But this difference in technique and in the variety of plants that can be produced does not make genetic engineering inherently more risky than other forms of genetic modification. Again, it is worth quoting the findings of the National Research Council:

> All evidence evaluated to date indicates that unexpected and unintended compositional changes arise with all forms of genetic modification, including genetic engineering. Whether such compositional changes result in unintended health effects is dependent upon the nature of the substance altered and the biological consequences of the compounds. *To date, no adverse health effects attributed to genetic engineering have been documented in the human population.* (Emphasis added)[13]

I emphasized the last sentence because it is worth emphasizing. No one has suffered any ill effects whatsoever from food because it was genetically engineered. This is not because genetically engineered plants are rare. To the contrary, most of the soybeans grown in the United States and over half of the corn have been genetically engineered. Several dozen different crop varieties have been genetically engineered and are in widespread use, including potatoes, tomatoes, and squash.[14] For over a decade, millions of Americans have been eating food derived from genetically engineered plants and have experienced no health problems as a result. As journalist Ronald Bailey notes, "no one has gotten so much as a sniffle, a stomach ache, or a rash because of biotech foods."[15]

This record is especially impressive when one realizes that much of the food we eat does prove to be unsafe, usually as the result of pathogens introduced through contamination (for example, through use of unsanitary water, improper food storage, and inappropriate food handling by workers). Every year 76 million Americans become ill from food-borne illnesses, resulting in 325,000 hospitalizations and 5,000 deaths. Not one of these illnesses is attributable to genetic engineering.[16] If antibiotech activists were motivated by a rational concern for food safety, they would push strenuously for more food inspectors as well as improved training for workers who transport or handle food, instead of agitating against genetically engineered food.

Why then the profound concern bordering on hysteria about the safety of genetically engineered food? To the extent this overwrought concern has any basis in fact it seems to be a reflection of the capability of genetic engineering to produce a greater range of organisms than other methods of genetic modification, as opposed to its actual track record. Genetically engineered plants can have combinations of genes impossible to achieve through standard crossbreeding. As indicated, one way genetic engineering differs from other forms of genetic modification is that DNA from many different organisms—completely unrelated to the plant to be modified—can be transferred to the plant. One of the arguable successes of genetic engineering has been the introduction of genes from the soil bacterium *Bacillus thuringiensis* into several different plants, including corn and cotton. This bacterium produces proteins harmful to some insects when ingested, including leaf-eating caterpillars, which are a major threat to crops. The safety and effectiveness of *Bacillus thuringiensis* (hereafter "Bt") as a pesticide source has been known for decades; indeed, organic farmers use Bt sprays and powders on their crops. (It is considered "natural" since it is derived from an organism.) Scientists developed Bt corn and Bt cotton by transferring genes from the bacterium—a transfer that would not have taken place without the techniques of genetic engineering—and these plants have been successfully cultivated, reducing the need for pesticides that otherwise would have been used. There is no need for these pesticides because the genetically engineered plant produces its own pesticide.

Of course, some of those familiar with Bt corn will argue that the history of this new plant illustrates the hazards of genetically engineered plants, in particular how such plants, because of their novel composition, can present unanticipated risks. In 1995 the Environmental Protection Agency (EPA) approved the marketing of Bt corn. A few years later, Dr. John Losey and some other researchers decided to test the effects of Bt corn on monarch butterfly larvae, after observing that milkweed could be found in and around many cornfields. Monarch larvae eat milkweed. In 1999, Losey's team published findings from a laboratory experiment indicating that monarch larvae that ate milkweed dusted with Bt corn pollen died at an alarming rate.[17] Roughly 44 percent of the larvae that ate the milkweed with Bt corn pollen died. Immediately there was a hue and cry about the dangers of genetically engineered food, with front-page headlines in the *New York Times* and elsewhere about butterfly-killing corn.[18] However, subsequent studies carried out by the National Academy of Sciences established there was little actual risk to the monarch larvae from BT corn pollen.[19] In real world field conditions, as opposed to the setting of a lab, the pollen from most Bt corn is not sufficiently concentrated to pose a threat to monarch larvae.

Both opponents and proponents of genetically engineered plants drew on the Bt corn–monarch butterfly studies for support. Opponents argued the episode illustrated the dangers of unanticipated risks from genetically engineered plants; proponents argued that, at the end of the day, all we had was another needless scare about the alleged hazards of genetically engineered plants.

Both sides had a point. Bt corn did prove to be safe, but it is also true that the effect on monarch larvae was not entirely anticipated.* Moreover, although most Bt corn pollen was not harmful to monarch larvae under actual field conditions, one type of Bt corn (known as Event 176) was capable of harming larvae in the field because of the high level of toxins in its pollen. (Event 176 has been withdrawn from use.)

There are a number of hazards routinely mentioned by opponents of genetically engineered plants, including unanticipated harmful effects on nontarget populations (e.g., butterflies as opposed to pests); produce that contains unforeseen allergens or toxins; and gene "flow" or "drift" through which the genes of transgenic plants are transferred to other plants, perhaps lending them characteristics that we would find undesirable (e.g., a weed that becomes resistant to herbicides). A risk related to

---

* I say "not entirely" because it was understood that Bt corn could kill the larvae of many insects. But the EPA had not realized that monarch butterflies were prevalent in many cornfields.

gene flow is the "contamination" of conventional produce with genetically engineered produce, which may or may not result from gene transfer. (It could result from the mixing of these products at some stage of their processing.) These are all within the realm of realistic possibilities. Some other projected consequences of genetic engineering will remain confined to the realm of science fiction. Despite the theoretical possibility of transferring sufficiently large numbers of human genes into plants to provide plants with human qualities, there is no scientific, commercial, or therapeutic incentive to do so, nor is it likely that such an organism would be viable, even if, somehow, a scientist were able to carry out such gene transfers. Brendan Sullivan, a lawyer representing Oliver North during the hearings on the Iran-Contra scandal, famously protested when he was instructed to keep quiet, "I am not a potted plant." The concern that a contrary assertion may one day emerge from the parlor palm in your office is a fear unlikely to be realized.

But the realistic possibilities at least arguably provide a sufficient basis for concern, if, in fact, there is no reliable way to regulate genetically engineered plants to avoid potential harmful effects. Is there? And what guidelines should we use in regulating genetically engineered plants?

With this background, we are now ready to address the central issue in this chapter, namely, how genetically engineered plants should be regulated. In particular, should we utilize the so-called "precautionary principle"? The precautionary principle, as formulated at the Wingspread conference,* states, in pertinent part, that:

> When an activity raises threats of harm to human health or the environment, precautionary measures should be taken even if some cause and effect relationships are not fully established scientifically. In this context the proponent of an activity, rather than the public, should bear the burden of proof.[20]

Effectively, pursuant to the precautionary principle, genetically engineered plants should not be approved unless they can be proven to be

---

* The Wingspread conference was a conference of environmentalists held in Wisconsin in 1998.

safe—both with respect to scientifically supported concerns *and* concerns that lack sound scientific support. Is this the method we should use for approval of genetically engineered plants, or should we apply some variant of standard risk assessment? This may sound like a very dry, academic question, but a lot is at stake. There is no doubt that application of the precautionary principle would seriously impede development of genetically engineered plants. On the other hand, it would likely protect us against unforeseen hazards from such plants.

Choosing the appropriate method of regulation has consequences not only for United States, but for countries around the world, and for global trade. For several years, the European Union had a moratorium on approval of genetically engineered plants based on the precautionary principle. This policy also limited imports of genetically engineered food products. After trade protests from the United States and other countries, the moratorium was lifted, but in its place there are stiff labeling and traceability requirements that add significant costs to trade in biotech products and effectively prevent global marketing of such products. Tracing the provenance of a particular food product and ensuring that conventional produce does not contain even trace amounts of genetically engineered produce is a very demanding requirement—and one that probably cannot be met 100 percent of the time. And, as I will discuss below, the consequences of even minimal blending of genetically engineered produce with conventional produce can be financially devastating for farmers and food processors given the paranoia surrounding genetically engineered food. For those who believe that genetically engineered plants can reduce our dependence on pesticides and herbicides, as well as increasing food production dramatically, burdensome restrictions on genetically engineered plants and produce are as shortsighted and harmful as they are unscientific. We will be sacrificing significant benefits based on unfounded fears. Although it has not received the same level of attention as the controversies over assistance in dying and stem cell research, the dispute over regulation of genetically engineered plants and produce is one of the most significant bioethical controversies of our time.

APPROPRIATE RISK ASSESSMENT

For the most part, as discussed in chapter 3, we encourage and respect self-determination. Even for choices we may regard as unreasonable, we tend to allow individuals to make their own decisions. This includes decisions about activities that expose a person to a risk of injury. You are free to skydive, climb mountains, and run with the bulls.

We also tend to allow individuals to engage in their own risk assessment. True, if a person is undergoing a risky activity with the help of a commercial enterprise, that enterprise will typically caution the person about the risks involved and perhaps require an express acknowledgment that the person has been informed of the risks—but that is less to dissuade the person than to protect the enterprise from litigation.

We encourage and respect self-determination for persons with risk-averse personalities as well. Typically, no one forces you to fly on a plane, try sushi, go to a party, drive after dark, or even read a book on bioethics. Many individuals decline to engage in activities that most of us find harmless, if not beneficial. If these individuals want to be "safe not sorry" and deprive themselves of a benefit because of an irrational concern, it's their business.

In addition to respect for self-determination, one reason we tend to allow individuals to decide for themselves whether to engage in certain activities is the difficulty in assessing the potential benefits and harms from activities that affect exclusively or principally the person engaging in those activities. Statistical studies will inform us that rock climbing is a fairly risky activity, roughly twelve times more risky than recreational boating, but the subjective pleasure one might obtain from rock climbing would be difficult for us to measure.[21]

So far, I have been making descriptive statements. I will now turn them into normative assertions. We should allow adults great latitude in terms of the self-regarding activities in which they choose to engage or refrain from engaging. As discussed in chapter 3, one element of our contemporary common morality is the belief that self-determination, all other things being equal, is valuable. Although it is more valuable with respect to critical life choices, it is also of some value even with respect to relatively inconsequential life choices. Moreover, the alternative, as indicated, would be for others to decide for us how beneficial we find rock climbing or how horrifying we find the prospect of going to a picnic. Such decisions are best left to the individual. This is not because the individual in question will always engage in the most rational risk assessment. To the contrary, an individual with strong preferences or aversions likely will place weight on considerations that others might find unwarranted if not incomprehensible, but that is one reason the choice should be left to the individual.

Of course, many, if not most, of our decisions are not self-regarding. Moreover, if we are discussing what policies legislatures or government agencies should adopt, then by definition these are decisions that are supposed to reflect the interests of the community. I went through the detour of discussing choices that are self-regarding principally to contrast them with choices that are made for and on behalf of the community. Here, we should place a premium on rationality. Indeed, with respect to public policy we have an obligation to make rational risk assessments. If I have a personal aversion to genetically engineered foods for whatever reason (or lack of reason), and this aversion leads me to buy organic, that's a matter of indifference as far as the general public is concerned. However, if I am advocating a policy that will affect the access that others have to genetically engineered plants and produce, I have an obligation to base my advocacy on sound science and objective considerations.

There is a vast literature on risk assessment, and in some contexts a discussion of risk assessment is necessarily very nuanced. Fortunately, here we can forego the nuances because the contrast between the precautionary principle and alternative regulatory principles is stark. One fundamental flaw of the precautionary principle is that it fails to factor in the benefits of a new activity (or, stated another way, the risks of not allowing the new activity). Risk analysis should be comparative. Consider vaccines. The introduction of polio, measles, DPT, and similar vaccines represented a tremendous advance in public health. For example, from approximately twenty thousand cases of paralytic polio a year prior to the introduction of the polio vaccine, the numbers dropped to a handful within a couple of decades. Similarly, measles killed several hundred people per year in the 1940s, but was virtually eliminated as a cause of death in the United States after introduction of the vaccine.[22] One reason the vaccines were so successful is that they were made mandatory for schoolchildren. These measures were not without risk. In fact, some children suffered serious adverse effects from the vaccines.

Moreover, scientists understood prior to the introduction of the vaccines that there was a risk a number of those vaccinated could become seriously ill or die. How great a risk was not known for sure. The polio vaccine was especially controversial because it was developed from a virulent poliovirus, and some members of the committee that had to approve the field trials were reluctant to give their approval because of the element of uncertainty. At the end of the day, however, approval was given principally because of the concern that a delay in introducing the vaccine would harm thousands.[23] Thus, a comparative risk analysis, carried out under conditions where the exact magnitude of the risk was unknown, persuaded the scientists to allow the field trials to proceed.

If the scientists had used the precautionary principle in determining whether to allow the field trials to proceed, these trials never would have taken place. There was simply no way the scientists could have proven the vaccines were safe. In fact, if "safe" means causing no harm whatsoever, then the vaccines were not and are not safe. Nonetheless, the benefits from the vaccines have far outweighed the harm they have caused.

No new technology has ever been risk free. The first wheeled vehicles exposed humans to risks that were novel and there undoubtedly were many injuries and deaths that could have been avoided had humans continued to confine themselves to foot power. Plus, think of the harm to the environment that resulted from the use of wheeled vehicles—harms that increased exponentially as we transitioned from simple carts to vehicles powered by internal combustion engines. On the other hand, it is undeniable that faster transportation also saved many lives in addition to bringing innumerable other benefits. Here again, implementation of the precautionary principle would have brought a halt to progress and we would have remained permanent pedestrians.

In chapter 1, I made a passing, humorous reference to objections that might have been raised by our ancestors with respect to the domestication of animals. The reality is that domestication was risky, although our ancestors certainly had no reliable way of calculating the risk, and not simply because they neither understood the role of pathogens in causing disease nor the biological processes through which animal pathogens could mutate and infect humans. Given the radically novel nature of this innovation, even if they had such knowledge, there was no means of making a precise calculation of the risks involved. Had the precautionary principle prevailed in ancient cultures, animals would never have become domesticated.

These examples highlight another problem with the precautionary principle. Not only does the principle fail to give appropriate consideration to the benefits that may result from an innovation, but the principle systematically exaggerates the weight to be given to the risks of a new activity. Under conditions of uncertainty, the default option is always to resist innovation. But why should we assume that the risks always outweigh the benefits when a precise calculation of risks and harms cannot be made? The precautionary principle effectively functions as a pseudoscientific validation of the bias for the status quo that is felt by many. In other words, many people instinctively prefer to maintain current conditions.[24] But this bias for the status quo is irrational. At the individual level, this is not much of a problem (except for the individuals involved), but at the level of policy a bias for maintaining the status quo is unjustifiable and harmful; regulation should be based on appropriate risk assessment, not prejudice against innovation.

Novelty does complicate the assessment of risk. There's no doubt about that. But it complicates the assessment of projected benefits as well as projected harms. The laser (an acronym derived from light amplification by stimulated emission of radiation) is a recent example of a technology whose potential applications were not recognized at the time it was first developed in the late 1950s. Of course, it is now an indispensable tool in manufacturing, telecommunications, medicine, home entertainment, and many other areas of contemporary life. However, shortly after its development, the public was focused on its potential harmful consequences, in particular, its use as a possible "death ray." One of its key developers, Theodore Maiman, was asked at parties whether he regretted inventing this death ray.[25]

Not every new technology, of course, will provide us with as many unexpected benefits as the laser. It would be irresponsible to allow a new technology to be implemented based on the fanciful hope that it will produce wonderful unexpected benefits, just as it would be incorrect to ban new technology based on the unfounded fear that it will produce serious unexpected harms. But the rational conclusion from the difficulty of predicting and comparing unforeseeable benefits and unforeseeable harms is that these unforeseeable benefits and harms should cancel each other out. Generally, they should be ignored for purposes of risk assessment. Instead, the precautionary principle instructs us to take into account unforeseeable harms, while dismissing the significance of unforeseeable benefits.

Despite the foregoing critique of the precautionary principle, I recognize that many may still prefer the precautionary principle if they believe our only choices for regulating genetically engineered plants are either the precautionary principle or an attitude of "anything goes." Fortunately, we can make realistic projections about both the potential harms and benefits of genetically engineered plants; in addition we can impose appropriate restrictions on the development of genetically engineered plants that should safeguard us against catastrophic consequences.

To begin, the gene transfer that takes place in creating new biotech plants is based on evidence about what that gene, or set of genes, does. Scientists do not transfer willy-nilly genes from one organism to another. Such a method would be an extremely inefficient use of resources. Instead, as the Bt corn example illustrates, scientists isolate the genes that produce the intended effects and then splice those genes into the plants that they want to manifest these effects.* The precision of genetic engineering is

*As anyone with even the most rudimentary science background will recognize, this is a simplification of a very complicated process. Among other things, genes do not work independently. They are regulated by other genes. Determining how to make genes function in the appropriate way is perhaps the most challenging aspect of recombinant DNA technology. However, this simplification does not affect the point I am making here.

one of the advantages that this technique has over other forms of genetic modification. Accordingly, there is a reliable—albeit not infallible—basis for predicting how the new plant will interact with the environment and humans.

In addition, of course, once the plant is grown, and produce from the plant is obtained, compositional testing of both the plant and the produce can take place to determine whether or to what extent the composition of the modified plant and produce differs from its "unmodified" comparator. In other words, we can analyze the composition of biotech rice to see how it differs from standard rice, and from these compositional differences we can project how the plant will interact with the environment and how the produce will affect those who consume it. As we have already noted, much has been made of the possibility of unanticipated effects resulting from genetic engineering. It is undeniable that unanticipated effects can occur—they can occur with conventional plant-breeding techniques as well. However, rigorous compositional analysis should reveal whether there has been an unanticipated change in the chemical makeup of the plant. Furthermore, we have enough background science to make reliable predictions about the safety of plants and foods based on this compositional analysis. We already know much about substances that constitute essential nutrients for humans as well as substances that are toxic to humans. Much concern has been raised about allergens that may be found in genetically engineered produce. But just as we have considerable knowledge about nutrients and toxins, we also understand which substances are likely to be allergens (eight types of food account for 90 percent of known allergens), so compositional testing of genetically engineered produce should reveal whether an allergen is present.

Admittedly, compositional analysis will not necessarily reveal exactly what characteristics the modified plant will manifest. Among other reasons, gene expression (what the gene does) is the result not only of the genetic composition of the organism but also of the organism's interaction with the environment. Moreover, determining how a plant engineered to produce pesticides might affect nontarget populations cannot be firmly established until those nontarget populations are exposed to the plant. Accordingly, in addition to lab testing, field testing for some modified plants may be appropriate.

In some instances, monitoring of the plant and the resulting food product after it has been approved should also be required. This is often referred to as "postcommercialization" or "postmarket" surveillance. This is one area where regulation of modified plants definitely could be strengthened. Postmarket surveillance is an accepted procedure for many drugs but has not been widely used or required for genetically modified plants. As the National Academy of Sciences has pointed out, although postmarket surveillance of modified plants may not be warranted when "compositional comparisons of a new GM crop or food . . . with its conventional counterpart indicate they are compositionally very similar," postmarket surveillance may be needed in some cases, such as when the modified plant or food product has "the potential to alter dietary intake levels (e.g., elevated levels of key nutrients)."[26]

I have intentionally used the term "modified" plants because there is no reason to exempt plants modified through conventional breeding or mutagenesis from compositional analysis or even, in some cases, field testing or postmarket surveillance. We have already discussed how plants modified by conventional breeding or mutagenesis can produce unanticipated effects, such as the allergic reactions caused by kiwi. As indicated, a reasonable argument can be made that testing of modified plants can be tightened and improved. (It would certainly help if the EPA and other regulatory agencies had a more informed understanding of the environment in which a modified plant will be cultivated—as the example of Bt corn and the monarch butterfly illustrates.) But there is no scientifically valid reason to impose testing requirements *only* on genetically engineered plants as opposed to plants modified by other means. The greater range of modifications available through genetic engineering does not make them inherently more dangerous.

Appropriate testing should diminish the risk of harm from allergens or adverse effects on nontarget populations. But what about genetic "flow," that is, the transfer of genes from genetically engineered plants into other plants? First, assuming there is no significant adverse impact on biodiversity, it is not clear whether gene flow would always be a bad thing. If a tomato that was engineered to be more resistant to pests cross-bred with standard tomatoes, wherein lies the harm? And with respect to biodiversity, there is no reason to suspect that genetically engineered plants will greatly reduce the number of plant variants and species, which, in any event, have already been reduced through conventional agriculture.[27] This leads to the next point. Genetic transfer occurs now between plants modified through other means and related species. Again, there is no reason in principle to be concerned only with genetically engineered plants as opposed to plants that have been modified through other means.

So gene flow is not necessarily a bad thing nor is it something unique to genetically engineered plants. That said, gene flow from genetically engineered plants does pose a couple of risks—one based on real biological consequences, one largely the result of irrational fear that, nonetheless, results in real financial consequences. Perhaps the primary concern raised with respect to gene flow has been the worry that plants engineered to be herbicide resistant (which has obvious advantages for farmers who use herbicides) will transfer this herbicide-resistant trait to weeds, creating a species of uncontrollable "super weeds." Such an occurrence is theoretically possible, although no super weeds have been created to date. Moreover, there are reasons to believe this risk, although not nonexistent, is not significant. First, herbicide-resistant plants, in particular, canola, have been created through conventional breeding techniques and have been in use for about two decades. No super weeds have been produced, even though canola is one commercial crop that has sexually compatible wild relatives in North America.[28] This leads to the next point. In the United States, most major commodity crops are not native and, therefore, do not have wild relatives. This greatly reduces the likelihood of harmful gene transfer. This fact, although important given the role of the United States as a major food producer, does not, of course, suggest we should not be concerned about cultivation of genetically engineered crops in countries that do have native weeds related to such crops.

However, even if herbicide-resistant genes were transferred to weeds, this trait would not persist and become dominant unless the weeds were subject to selection pressure. In other words, they would have to be sprayed repeatedly with the same herbicide to give the weeds with the herbicide-resistant gene a competitive survival advantage. Intelligent use of herbicides should prevent the occurrence of such a scenario.

Finally, there are technologies to prevent gene flow. One is so-called "terminator" technology, by which seeds can be made sterile. Sterile seeds would prevent bioengineered traits from being transferred into weeds through cross-pollination. However, in one of the ironies of the controversies over genetically engineered plants and foods, when Monsanto announced tentative plans to purchase the seed company (Delta and Pine Land) that had developed this technology, a storm of protest ensued. One consequence of sterile seeds is that farmers who want to use engineered seeds would have to buy new seeds for each growing season. At least for many farmers in North America, this would not be a major change in practice, as they have been buying seeds annually for many years. (The progeny of high-yielding hybrid seeds do not reliably manifest the desirable characteristics of the parental stock—hence the need for annual seed purchases.) Nonetheless, those opposed to genetically engineered plants accused Monsanto of scheming to make farmers dependent on their products, allowing them to dominate world food production. The public relations campaign of the anti-biotech activists was successful and Monsanto eventually announced it could not market terminator technology. It also backed out of its deal to purchase Delta and Pine Land (which led to litigation).[29]

By protesting the marketing of terminator technology, the anti-biotech crowd was also blocking one of the methods that would eliminate one of the allegedly greatest risks of genetically engineered plants, namely, the spread of genes from these plants to their wild cousins. They were blocking the solution to the very problem they identified as critical. This campaign against terminator technology was very revealing of the motives of (some of) those opposed to genetic engineering—they do not want it under any circumstance. Genetically engineered plants allegedly pose a risk of genetic flow, but solving this problem by creating plants with sterile seeds allegedly gives too much power to the corporations that market these seeds. Against this Catch-22, you can't win. The result: no genetically engineered plants.[30]

Fortunately, there is another means of reducing the risk of gene flow, and that is by introducing transferred genes into the chloroplasts of plants. (Chloroplasts are the cellular structures that help convert sunlight into energy in many plants.) Because chloroplasts are not found in the pollen of most crop plants, this method of gene transfer should greatly reduce gene flow. Indeed, scientific studies have confirmed that this technique virtually eliminates gene flow from genetically engineered plants to related species. (It does not eliminate the possibility of gene flow completely because accidental distribution of seed, for example, in harvesting or transporting plants, can result in spreading seed from genetically engineered plants to related species.)[31]

All things considered, the risk of gene flow creating a species of super weeds or some other noxious plant is minimal. There are technologies to prevent or diminish gene flow, and even if we choose not to use these technologies, several other factors would have to come into play before any super weed became a reality.

In addition to gene flow that might have real biological consequences, there is gene flow that causes no threat of biological harm, but results in financial harm to food producers—because of the fear of genetically engineered food and the demands of some countries, principally those in the European Union, that conventional food be strictly segregated from genetically engineered food. There have been a couple of incidents in the last decade in which conventional produce was found to contain minimal amounts of genetically engineered produce. The last significant incident occurred in 2006, when shipments of rice to France were found to contain trace amounts of LLRICE 601, or Liberty Link rice, a rice engineered to be herbicide tolerant. To make matters worse, LLRICE 601 had not been approved for human consumption. In the resulting uproar, rice exports to Europe were halted and rice prices tumbled. Many rice farmers took a financial hit.[32]

But what exactly was the harm here and who was responsible? First, although I have stated that the trace amounts of LLRICE 601 found in conventional rice may have resulted from gene flow, as of the date I am writing this, no one knows for sure. The most widely accepted working hypothesis is that at least one experimental field of LLRICE 601 had been planted in proximity to a field of conventional rice, resulting in some cross-pollination. But it's also possible that there was some inadvertent mixing of LLRICE 601 with conventional rice, without any gene flow. Since the Department of Agriculture has now ended its investigation, we may never have a definitive answer.[33] However, for the sake of argument, let us assume that the LLRICE 601 that was found among conventional rice resulted from gene flow.

The first thing to notice is that this cross-pollination did not result in the supplanting of conventional rice with a genetically engineered variant, nor did it create some sort of noxious rice variant. Small amounts of the biotech rice were found among conventional rice. Period. Of course, there was much ado about the fact that the biotech rice had not been approved for human consumption, but that was because Bayer, the owner of LLRICE 601, *had never applied for approval as it never intended to market it.* Once it became known that LLRICE 601 could be found in the conventional food supply, Bayer petitioned the Department of Agriculture for approval, which quickly determined it was safe for human consumption and required no special regulation.[34] Unfortunately, the European Union still refused to accept American rice, despite the absence of any scientific evidence showing the rice was unsafe.

Those who are opposed to genetically engineered plants and foods have argued this latest episode confirms that biotech plants threaten our food supply. It shows no such thing. Admittedly, unless terminator technology is introduced, there will be incidents in which genetically engineered plants cross-pollinate with conventional plants. Even with such technology, there probably will be some occasions of inadvertent mixing during the processing of food from the different types of plants. Food processing—whether it's a question of having workers wash their hands or segregate produce appropriately—ultimately depends on humans, who are not reliable and conscientious 100 percent of the time. But the relevant question is whether this blending of genetically engineered plants with conventional plants poses any risk to us. There is no evidence it does.

Of course, the farmers who could no longer export their rice to Europe suffered real, quantifiable financial harm, but it seems to me that this is the fault of the irresponsible policies of the bureaucrats in the European Union and their irrational supporters. Should we ban biotech plants simply because of the ignorance and prejudice of some? Do you remember the rumors about a decade ago regarding some Mexican beer being contaminated with urine? These rumors had no basis in fact, but there were many gullible persons who believed them and they did have a severe economic impact on a couple of Mexican exporters.[35] Should the United States have given in to popular prejudice and banned Mexican beer? I think not, nor do I think we should prohibit genetically engineered plants to render our agricultural policies consistent with the irrational fears held by many—whether they are located in Brussels or elsewhere.

To sum up, all the principal realistic risks arising from genetically engineered plants—adverse effects on nontarget populations, production of foods with allergens or other harmful substances, and the creation of a super weed or some other noxious plant through the transfer of traits from genetically engineered plants—can be reduced to the point of insignificance by a combination of appropriate regulation and cultivation techniques.

However, we have examined only one side of the ledger. If genetic engineering provided no benefits, then we might prohibit genetically engineered plants because even a small risk of harm outweighs an absence of benefits. But genetically engineered plants do provide measurable benefits already, particularly in the area of reducing use of pesticides and in increasing crop yields. To take just one example, a study of six cotton-growing states in the period between 1995 and 1999 found that the introduction of Bt cotton had reduced pesticide use by about 14 percent. More effective pesticides should also produce greater yields, and this also was confirmed by an analysis of the 1999 cotton harvest, which revealed that use of Bt cotton resulted in an average yield increase of 9 percent.[36]

Another recent study confirmed that in 2005 alone, the gain in production from American Bt crops was valued at $553 million.[37]

In addition, reduction in the use of pesticides not only has obvious environmental benefits, but it has a direct connection to a decline in deaths and illnesses suffered by agricultural workers in developing nations, where pesticides are often applied by hand. For example, after Bt cotton was introduced in China, "farmers who planted only Bt varieties reported just one-sixth as many pesticide poisonings per capita as those who planted only conventional cotton."[38]

Herbicide-resistant plants also have provided significant benefits. First, herbicide-resistant plants lead to more effective and efficient use of herbicides. For fields planted with herbicide-resistant plants, there has been a net reduction in the number of herbicide applications, with a savings in weed control costs of hundreds of millions of dollars.[39] Perhaps more important, herbicide-resistant plants have allowed farmers to use more benign herbicides that degrade rapidly. Monsanto's Roundup Ready soybean has become immensely popular with farmers and now accounts for nearly 90 percent of all soybean acreage in the United States.[40] As indicated by its name, these soybeans can resist exposure to the Monsanto-produced herbicide Roundup.* Roundup is considered preferable to many other herbicides because it is less toxic to mammals, birds, and fish and degrades twice as quickly as other herbicides.[41] Finally, because herbicide-resistant plants require less tillage to control weeds, they reduce topsoil erosion and soil runoff. A major increase in no-till acreage started with the introduction of herbicide-resistant plants in 1996 and no-till acreage has continued to expand as more herbicide-resistant plants are adopted by farmers.[42]

The reduction in pesticide use, the increase in crop yields, and the use of more environmentally friendly cultivation methods as a result of

---

* Yes, Monsanto benefits twice from farmers turning to Roundup Ready soybean, first by selling the soybean and then by selling the herbicide. But so what? As the reader may have gathered, one of the underlying concerns of some of those opposed to genetically engineered plants and foods is that these products often are developed by for-profit corporations. By itself, this fact is irrelevant to a rational risk assessment. Corporations are neither our friends nor our enemies, and food should not be exempt from regulation nor subject to overregulation merely because a corporation has produced it.

herbicide-resistant plants would be sufficient by themselves to conclude that genetically engineered plants have provided substantial benefits. However, there are other benefits as well, including the development of virus-resistant papaya and squash. The papaya mosaic virus devastated papaya cultivation in Hawaii in the 1990s. Hawaiian papaya farmers are now back in business and virus-resistant papaya currently comprises 55 percent of the papaya crop in the United States.[43] The ability to immunize crops from infectious diseases is of immense significance. Some readers may recall that the devastating famine in Ireland in the 1840s resulted from potato blight, a disease caused by a fungus. (Obviously, a major contributing factor was the dependence of many Irish on a single crop.) Use of genetically engineered crops that are immune to infectious diseases should safeguard our crops and prevent similar disease-related food shortages.

Another benefit—the creation and widespread dissemination of more nutritious food—is at this stage more of a work in progress than a reality, but significant progress has been made. The most significant, and controversial, development is the creation of Golden Rice, a genetically engineered rice variant that is rich in beta-carotene in its endosperm, the portion of rice that is typically consumed.* Beta-carotene is a precursor of vitamin A. The development of Golden Rice was an amazing technological accomplishment, requiring the transfer of multiple genes from daffodils and bacteria. A major impetus for the development of this variant is that vitamin A deficiency is a leading cause of blindness and/or a contributing factor in the deaths of millions each year, especially in developing countries. Golden Rice, if properly cultivated and distributed, could save the sight and lives of many, with recent estimates of between five thousand and forty thousand lives saved in India alone each year.[44]

Golden Rice was first developed in the late 1990s. So why has it not been widely cultivated and distributed yet? Several reasons—and here is one instance where both the proponents and the opponents of genetically engineered plants and produce share some blame.

---

* Golden Rice has no commercial or biological connection to LLRICE 601, discussed a few pages earlier.

Some proponents of Golden Rice hyped it excessively, suggesting that it would resolve the problem of vitamin A deficiency in developing countries. However, Greenpeace—an organization adamantly opposed to genetically engineered plants and produce—decided to put some of this puffery to the test. Greenpeace calculated that children would need to eat about six pounds of rice per day to obtain the minimum amount of vitamin A needed to stave off illness. This is a ludicrously large amount of rice. Faced with these embarrassing counterarguments, proponents of Golden Rice had to backtrack, admitting that Golden Rice would not provide the entire solution to vitamin A deficiency. They nonetheless insisted—quite reasonably—that Golden Rice would provide part of the solution, as a complement to other dietary components. Unfortunately, the damage had been done and the assault by Greenpeace was effective in eroding at least some of the support for Golden Rice and delaying its cultivation in India and other counties.[45]

This setback was unfortunate because Golden Rice is beneficial. Moreover, a second strain of Golden Rice has been developed that contains considerably more beta-carotene (the daffodil gene was replaced with an equivalent gene from corn) and, therefore, can deliver much more vitamin A per bowl of rice. A study of Golden Rice 2 has demonstrated it would be very effective in reducing vitamin A deficiency, "although it is no panacea in the fight against malnutrition."[46] Among other obstacles to be overcome is the need to have the relevant traits incorporated into rice strains that are hardy and sustainable under conditions prevailing in developing countries.

But the biggest set of obstacles that remains are the regulatory hurdles put in place *solely* because Golden Rice has been genetically engineered. The principal developer of Golden Rice has estimated that its adoption has been delayed by at least seven years because of the extreme precautionary reactions to genetically engineered products.[47] These reactions lack scientific justification. It was entirely appropriate for Greenpeace to subject the hyped claims for Golden Rice to empirical testing. It is entirely inappropriate for Greenpeace to argue for bans on genetically engineered plants and foods based on a blend of pseudoscience and anticorporate rhetoric.[48] While the development and adoption of plants such as Golden Rice have been delayed, hundreds of thousands are suffering due, in part, to the irresponsibility and ignorance of those who cannot separate science fiction from science fact.

To sum up, genetically engineered plants and foods are delivering substantial benefits now and, assuming they are not banned, it is highly likely they will provide additional benefits in the future, including crops that will supply some essential nutrients to those in need. I have intentionally omitted from my discussion of benefits those developments that are possible, maybe even probable, such as the development of foods from which allergens have been removed, crops with vastly increased yields, or biotech forests that will produce more paper or lumber with fewer trees. These have not been realized yet and may never be. Just as we deeply discounted speculative claims about some theoretical risks from genetically engineered plants, similarly we should discount projected but unrealized benefits—even though the latter seem more likely. But even confining ourselves to realized benefits, it is readily apparent they outweigh realistic risks that, as indicated, can be controlled through appropriate regulation and cultivation methods. On this policy issue, my conclusion stands on as firm a foundation as science can provide. If you disagree—well, fine. You can shake your head in disbelief as you share your mutated (but organic!) grapefruit with your friend Sasquatch.

DESSERT

Even though I have addressed the principal policy issue presented by this chapter, I cannot resist taking a quick bite out of another question, and that is the issue of labeling genetically engineered products. Even some who agree that the fears about genetically engineered foods lack scientific support believe these food products should be labeled out of respect for consumer choice and marketplace values.[49]

Although I am both a firm defender of self-determination and, as my wife can confirm, someone who actually reads the labels on packages, requiring labeling of products as genetically engineered would represent a capitulation to the forces of ignorance and irrationality. Moreover, to the extent mandatory labeling implies that genetically engineered food is not as safe as other food, the labeling would be misleading.

The regulatory forest on labeling is pretty thick, but I believe I can map out some key points without getting lost. The Food and Drug Administration (FDA) has the authority to require food labels to disclose material information, for example, the presence of an allergen or an additive. When it formulated policy on genetically engineered food in 1992, the FDA sensibly concluded that no labeling was required because "the method of development of a new plant variety . . . is [not] material information and would not usually be disclosed in labeling for the food."[50] Put simply, as a matter of science it is the chemical composition of the product that counts, *not* the way in which it was produced.

The argument that we should require labeling out of respect for consumer choice overlooks the fact that consumers can already avoid genetically engineered foods if they want by shopping organic. Organic producers are not shy about labeling their products. In fact, that was one of the primary motivations for pushing for official recognition of organic foods. More fundamentally, it is an imprudent, if not immoral, policy to encourage consumers to adhere to their prejudices. When AIDS first appeared on the scene, many feared that it could be spread through casual contact. That fear was quickly debunked scientifically, but undoubtedly many people still have that unfounded fear. Even without the fear of AIDS, some people still have an irrational aversion to gays. Should we require labels on food to disclose whether it was handled by gay workers? How about labels that disclose the astrological sign of the CEO of the company that markets the food? Such policies arguably would respect consumer choice and marketplace values.

As the reader can undoubtedly discern, these are rhetorical questions, but they illustrate the danger of requiring labeling based on prejudices or paranoia. Either the information is material or it is not. The strong consensus of the scientific community is that a food product derived from a genetically engineered plant is not inherently less safe. The FDA should assist us in making informed intelligent choices about how we feed ourselves; the FDA should not feed superstition.

This chapter has discussed plants and food that have been genetically engineered to produce desirable traits. Humans also can be modified to have desirable capacities and traits. The next chapter will discuss which policies we should adopt—if any—to regulate the use of pharmacological or genetic enhancements for humans.

NOTES

1. Union of Concerned Scientists, "Food and Environment: A Special Note to Organic Consumers," http://www.ucsusa.org/food_and_environment/genetic_engineering/note-to-organic-consumers.html (accessed August 18, 2007).

2. See Gregory Conko, "The Benefits of Biotech," *Regulation* (2003) 20–25. This article is available at http://www.cato.org/pubs/regulation/regv26n1/v26n1-4.pdf (accessed August 19, 2007). See also Lee M. Silver, *Challenging Nature: The Clash between Biotechnology and Spirituality* (New York: Harper Perennial, 2007), pp. 235–36.

3. See Organic Foods Production Act Provisions, Synthetic Substances Allowed for Use in Organic Crop Production, *Code of Federal Regulations*, title 7, sec. 205.601 (2007).

4. See Samuel Fromartz, *Organic, Inc.: Natural Foods and How They Grew* (New York: Harcourt, 2006), pp. 6–12. See also Silver, *Challenging Nature*, pp. 225–29.

5. Fromartz, *Organic, Inc.*, p. 9.

6. Organic Foods Production Act Provisions, Terms Defined, *Code of Federal Regulations*, title 7, sec. 205.2 (2007).

7. Gregory E. Pence, "Organic or Genetically Modified Food: Which Is Better?" in *The Ethics of Food: A Reader for the 21st Century*, ed. Gregory E. Pence (Lanham, MD: Rowman and Littlefield, 2002), pp. 116–22.

8. Helga Willer and Minou Yussefi, eds., *The World of Organic Agriculture:*

Statistics and Changing Trends, 9th ed. (Bonn, Germany: International Federation of Organic Agriculture Movements, 2007), p. 15. This report is available at http://orgprints.org/10506/01/willer-yussefi-2007-p1-44.pdf (accessed August 20, 2007).

9. Nina V. Federoff, "Agriculture: Prehistoric GM Corn," *Science* 302 (2003): 1158–59.

10. Committee on Identifying and Assessing Unintended Effects of Genetically Engineered Foods on Human Health, National Research Council, *Safety of Genetically Engineered Foods: Approaches to Assessing Unintended Health Effects* (Washington, DC: National Academies Press, 2004), p. ix.

11. Ibid., p. 5.

12. "Wide crosses" between unrelated species became possible in the last century through specialized cultivation techniques of hybrid plant embryos (which would otherwise be sterile). In fact, most of the bread wheat varieties grown in the United States are products of such wide crosses. Although discussion of such wide-cross plants is tangential to my main argument, it is important to be aware of such wide crosses when one hears zealots rant against the "unnatural" character of genetic engineering. As with mutagenesis, wide crosses have attracted scant attention from opponents of genetically engineered plants.

13. Ibid., p. 8.

14. Henry I. Miller and Gregory Conko, *The Frankenfood Myth: How Protest and Politics Threaten the Biotech Revolution* (Westport, CT: Praeger Publishers, 2004), p. 8.

15. Ronald Bailey, *Liberation Biology: The Scientific and Moral Case for the Biotech Revolution* (Amherst, NY: Prometheus Books, 2005), p. 190.

16. Centers for Disease Control and Prevention, *Foodborne Illness* (Washington, DC: CDC, 2005), p. 5. The report is available online at http://www.cdc.gov/ncidod/dbmd/diseaseinfo/foodborne_illness_FAQ.pdf (accessed August 26, 2007).

17. John E. Losey, Linda S. Rayor, and Maureen E. Carter, "Transgenic Pollen Harms Monarch Larvae," *Nature* 399 (1999): 214.

18. Carol Kaesuk Yoon, "Altered Corn May Imperil Butterfly, Researchers Say," *New York Times*, May 20, 1999, p. A1.

19. See, for example, Mark K. Sears et al., "Impact of Bt Corn Pollen on Monarch Butterfly Populations: A Risk Assessment," *Proceedings of the National Academy of Sciences* 98 (2001): 11937–42. There were actually six separate studies carried out by the National Academy of Sciences on this issue.

20. Carolyn Raffensperger and Joel Tickner, eds., *Protecting Public Health and the Environment: Implementing the Precautionary Principle* (Washington, DC: Island Press, 1999), pp. 353–54.

21. On the risks of rock climbing versus recreational boating, see Joshua T. Cohen and Peter J. Neumann, "What's More Dangerous, Your Aspirin or Your Car? Thinking Rationally about Drug Risks (and Benefits)," *Health Affairs* 26 (2007): 636–46, especially exhibit 4.

22. Centers for Disease Control and Prevention, "What Would Happen If We Stopped Vaccinations?" Available at http://www.cdc.gov/vaccines/vac-gen/whatifstop.htm (accessed September 19, 2007).

23. Marcia Meldrun, "'A Calculated Risk': The Salk Polio Vaccine Field Trials of 1954," *British Medical Journal* 317 (1998): 1233–36.

24. A classic article discussing status quo bias is William Samuelson and Richard Zeckhauser, "Status Quo Bias in Decision Making," *Journal of Risk and Uncertainty* 1 (1988): 7–59.

25. "Laser Inventor Dies at 79," photonics.com, May 9, 2007. Available at http://www.photonics.com/content/news/2007/May/9/87661.aspx (accessed September 29, 2007).

26. Committee on Identifying and Assessing Unintended Effects of Genetically Engineered Foods on Human Health, *Safety of Genetically Engineered Foods*, p. 183.

27. The concern with the elimination of biodiversity is based principally on the fear that this would make key crops susceptible to devastation by a disease, which could then create a serious food shortage.

28. Conko, "The Benefits of Biotech," p. 23.

29. My summary of the controversy over terminator technology draws on two separate sources, with interestingly different interpretations of the dispute: Bailey, *Liberation Biology*, pp. 201–204, and Marion Nestle, *Safe Food: Bacteria, Biotechnology, and Bioterrorism* (Berkeley: University of California Press, 2003), pp. 229–32.

30. Readers will be reminded of one of the arguments against assistance in dying: only the depressed will request assistance in dying and if you're depressed, you're not competent to request assistance in dying. The result: no assistance in dying.

31. Susan E. Scott and Mike J. Wilkinson, "Low Probability of Chloroplast Movement from Oilseed Rape (*Brassica napus*) into Wild *Brassica rapa*," *Nature Biotechnology* 17 (1999): 390–93.

32. For an overview of the controversy over LLRICE 601, see Marc Gunther, "Attack of the Mutant Rice," *Fortune*, July 9, 2007, pp. 74–83.

33. Rick Weiss, "Probe into Tainted Rice Ends," *Washington Post*, October 6, 2007, p. A2.

34. Animal and Plant Health Inspection Service, Department of Agriculture, "Bayer Crop Science; Extension of Determination of Nonregulated Status to Rice Genetically Engineered for Glufosinate Herbicide Tolerance," *Federal Register* 71 (December 4, 2006): 70360–62.

35. For an academic treatment of this urban legend, and some similar ones, see Gary Alan Fine, "Mercantile Legends and the World Economy: Dangerous Imports from the Third World," *Western Folklore* 48 (1989): 153–62.

36. Conko, "The Benefits of Biotech," p. 22.

37. Sujatha Sankula, *Quantification of the Impacts on US Agriculture of Biotechnology-Derived Crops Planted in 2005: Executive Summary* (Washington, DC: National Center for Food and Agricultural Policy, 2006), p. 7. This report is available at http://www.ncfap.org/whatwedo/pdf/2005biotechExecSummary.pdf (accessed September 30, 2007).

38. Conko, "The Benefits of Biotech," p. 22.

39. Janet E. Carpenter and Leonard P. Gianessi, *Agricultural Biotechnology: Updated Benefit Estimates* (Washington, DC: National Center for Food and Agricultural Policy, 2001), p. 2. This report is available at http://www.ncfap.org/reports/biotech/updatedbenefits.pdf (accessed September 30, 2007).

40. Sankula, *Quantification of the Impacts on US Agriculture of Biotechnology-Derived Crops Planted in 2005*, p. 6.

41. Bailey, *Liberation Biology*, p. 188.

42. Sankula, *Quantification of the Impacts on US Agriculture of Biotechnology-Derived Crops Planted in 2005*, p. 9.

43. Ibid., p. 6.

44. Alexander J. Stein, H. P. S. Sachdev, and Matin Quinn, "Potential Impact and Cost-Effectiveness of Golden Rice," *Nature Biotechnology* 24 (2006): 1200–1201.

45. Information for this paragraph is derived from two principal sources: Nestle, *Safe Food*, pp. 159–66, and Greenpeace International, "Golden Rice Is Fool's Gold," in *The Ethics of Food*, pp. 71–73.

46. Stein et al., "Potential Impact and Cost-Effectiveness of Golden Rice," p. 1201.

47. Ingo Potrykus, "Comments on the World Development Report 2008,"

available at http://www.goldenrice.org/Content4-Info/info7_actuality.html (accessed September 30, 2007).

48. The reader can judge for herself by consulting the Web site of Greenpeace International at http://www.greenpeace.org/international/campaigns/genetic-engineering.

49. Nestle, *Safe Food*, pp. 225–26.

50. Food and Drug Administration, "Statement of Policy: Foods Derived from New Plant Varieties," *Federal Register* 57 (May 29, 1992): 22991.

6.

"How Horrible! You've Become More Intelligent!" – Is Self-Improvement Wrong?

Most of us like to acquire new skills that will help us achieve our goals, even if these goals are something as simple as enjoying a new pastime. We want to learn to speak Chinese or handle a circular saw or play a piano. In addition to acquiring new skills, we also want to acquire new traits and capacities that will allow us to utilize our skills effectively. We want to become smarter or stronger or to possess the nimbleness, dexterity, and coordination of a good piano player.

The accuracy of these statements certainly is assumed by the multitude of companies that market products to us to help us acquire these skills, traits, and capacities. We spend billions of dollars each year on language tapes, piano lessons, athletic equipment, and so on. One strong selling point of many of these products is the speed at which they will allow us to acquire the desired skills, traits, and capacities. The ads scream out at us from television, radio, and magazines: Learn Chinese in three weeks! Lose that gut in two months! Five easy steps to understanding stocks! Our time is valuable and the faster we can acquire a skill, trait, or capacity, the better.*

One reason we find time valuable is our perception that life is short. Most of us have many goals that we'd like to pursue and seventy or eighty years do not seem sufficient. Hence our desire for techniques or

* With my usual caveat: all other things being equal. If the process of learning the skill or acquiring the capacity is pleasurable in itself, we may not want to rush things. I'll let the reader imagine his own examples.

substances that will help us prolong our lives, especially if they also promise to keep us fit and relatively vigorous. We guzzle down anti-oxidants and vitamin E, take synthetic hormones, practice yoga, engage in aerobics, dose ourselves with a blend of herbs, reduce our calorie intake—to name just a few of the less bizarre methods persons employ to extend their lives. There is an entire industry devoted to such anti-aging treatments and tactics.

Whatever we may think of the prudence of a person's pursuit of a skill or attempt to improve her capabilities—or efforts to retard aging—we typically do not condemn the person as immoral because of her activities, nor do we think her activities dehumanize us or threaten any significant adverse impact on society. Maybe you'd rather have your daughter studying physics than skateboarding, but the fact that she wants to improve herself in some way, as opposed to sitting in front of the television or connected to her iPod, is usually regarded as good thing. Self-improvement is usually considered something we should encourage. Some might even argue that we have a moral obligation to improve ourselves, at least in some ways.

However, the reaction of many becomes markedly different when the self-improvement under consideration results from some drug or a modification of one's genes that promises significant, rapid improvements in core capacities such as memory, intelligence, strength, endurance, or agility. What was praiseworthy now suddenly becomes morally suspect and, to some, frightening. We have government commissions study how these methods of self-improvement should be regulated, and some bioethicists and other scholars call upon us to "draw red lines" and ban self-improvement through pharmacological or genetic enhancements.[1]

In this chapter, I will address two related policy questions. Many have objected to pharmacological or genetic enhancements because of concerns about unequal access to these enhancements. Underlying these concerns is the not unreasonable belief that, in the absence of government regulation, access to enhancements—at least initially—may be limited to the wealthy. The wealthy will use their advantages in financial resources to acquire enhanced intellectual capacities, such as memory and analytical ability, and physical capacities, such as strength and endurance. These advantages, in turn, will translate into even greater success in obtaining and maintaining wealth, political power, and other desirable social goods, thus exacerbating existing inequalities. Recognition of the advantages conferred by enhanced capacities has led some to suggest that access to certain enhancements may have to be restricted if these enhancements threaten to impose "unjust limitations on opportunity to others."[2] Maxwell J. Mehlman, a bioethicist at Case Western Reserve University, is especially troubled by the prospect of germline genetic enhancement.* Mehlman has issued dire warnings about the emergence of a "genobility," a caste of enhanced humans who would rule over the rest of us.[3] He has called for a complete ban on germline genetic enhancement and has argued for severe restrictions on somatic genetic enhancements.[4]

But ensuring equal access to enhancements is not the only concern of those who object to the development and use of enhancements. Michael J. Sandel, a member of the President's Council on Bioethics, has stated, "The fundamental question is not how to assure equal access to enhancements but whether we should aspire to it."[5] Sandel sees enhancements as the product of a lamentable desire to master nature that endangers our sense of "giftedness," by which he means the fact that currently our traits and capacities are still largely outside of our control. He views what he characterizes as a lack of responsibility for shaping ourselves as a good thing: "One of the blessings of seeing ourselves as creatures of nature, God, or fortune is that we are not wholly responsible for the way we are."[6] The availability of genetic enhancements will increase "the burden we bear for the talents we have and the way we perform." There are others who regard enhancements as morally objectionable regardless of their effects on social equality. The ubiquitous bioconservative Leon Kass has objected to enhancements because they lessen the significance of human agency and human effort,[7] and George Annas opines that using enhancements to improve the capabilities of our children converts them into "manufactured products."[8]

---

* "Germline" (or "germ-line") genetic enhancements are enhancements that alter the DNA of gametes, and, therefore, can be passed on to subsequent generations. Enhancements that affect only the individual enhanced are often referred to as "somatic" enhancements, because they affect cells not used in reproduction, that is, somatic cells.

So we have two issues to consider. One is whether there is something wrong about using enhancements, whatever their effect on social equality. The other is what restrictions, if any, should be placed on enhancements because of the alleged threat they pose to social equality. I will address each in turn, beginning with the issue of whether enhancements are objectionable in and of themselves. Note that there are several issues relating to enhancements that I will not address, including their safety and whether it is appropriate to expend resources developing enhancements when there are arguably more pressing problems in the world, such as something as basic as ensuring access to clean drinking water for millions of unfortunate individuals. Safety is obviously an important consideration, but both proponents and opponents assume that some safe enhancements will eventually be available. I do not quarrel with the proposition that more attention should be paid to basic healthcare needs, but enhancements are going to be developed whatever reallocation of resources we make, in part because many enhancements will result from research into needed therapies.

Let me say something now about the state of enhancement technology, before proceeding to the policy discussion. Despite the deluge of articles and books addressing genetic or pharmacological enhancements, and ominous, if not apocalyptic, predictions about our "posthuman" future, we are not close to possessing the ability to design ourselves or our progeny. The fact is that direct application of biotechnology to enhance a wide range of specific human traits and capacities in a safe and predictable manner is not, for good or ill, a goal we are going to realize anytime soon.

Genetic enhancements are especially iffy. In the last chapter, we discussed the successful use of recombinant DNA technology with plants. Humans are a bit more difficult to work with than plants. In addition, when considering human enhancements, we are usually thinking of complex traits or capabilities, such as intelligence, memory, and so forth, that are not associated with any single gene, as opposed to the limited enhancements for genetically engineered plants. Furthermore, it is not just hypothetical Gene Sets A–D, which we may find are associated with many intelligence-related capacities, that we would have to modify; we probably would have to modify genes located in other areas of the person's genome, which may regulate the expression of the genes associated with intelligence. Genes interact in complicated patterns with other genes. Genetic modification is not like replacing a blown fuse. Despite some recent promising experiments with genetically altering certain brain receptors associated with memory formation in mice,[9] we are nowhere near being able to specify exactly which portions of a human's genome need altering to produce an enhancement as intricate and multifaceted as improved intelligence. And we are even farther away from knowing how to deliver the selected DNA package so that it is processed in the desired way. Attempts to use genetic interventions as therapies, that is, as a means to cure someone's disease or impairment, remain very risky, and there have been far more failures, including patient deaths, than successes.[10] As of now, the most widely used method of delivering therapeutic DNA is to use viruses, which penetrate the patient's targeted cells and deposit the therapeutic DNA. There are a number of problems with this method, including integrating the DNA into the patient's genome and immune and inflammatory responses, even though the diseases treated so far require relatively simple corrections to the patient's DNA. Identifying and delivering genes to enhance complex traits, as opposed to remedying some defect associated with a small set of genes, requires much more knowledge and precision in technique than we currently possess or are likely to possess in the near term. Indeed, some predict we may never have the ability to design humans to our genetic specifications.[11]

Pharmacological enhancements, that is, drugs as opposed to genetic modifications, are much closer to fruition. In fact, we already have some drugs that can enhance our traits and capacities, including one that has resulted in a revolutionary social transformation (more on that below). But even here we have not yet developed a large array of drugs that can safely and predictably deliver substantial beneficial changes to our traits or capacities. As indicated, there is at least one widely available drug that is safe and reliable and has enhanced a critical capacity for many. (Keep guessing which drug I am referring to for another few pages.) And the technology exists to enhance haphazardly a few other traits and capacities, such as equanimity, height, and the ability to concentrate. Prozac, growth hormone, and Ritalin can be and are used by those who are not clinically depressed, abnormally short, or suffering from ADHD (attention deficit and hyperactivity disorder) to enhance their self-confidence, stature, or ability to focus, respectively. Modafinil, a drug developed to treat narcolepsy, has also been used by some in recent years to suppress the need for sleep and to enhance certain discrete cognitive skills, such as visual pattern recognition memory and reaction time.[12] But not only is access to such drugs still relatively restricted (you need a very cooperative—if not unethical—physician to get a prescription for Ritalin or modafinil if you are not suffering from ADHD or a sleep disorder), but they do not constitute the magic pill that most people have in mind when they discuss the prospects for reshaping human nature. Drugs that would safely enhance complex traits and capacities such as intelligence, which obviously encompasses a range of cognitive skills and abilities, are not now available at any price.

Finally, through genetic screening and embryo selection or abortion, we have the ability to avoid some unwanted traits or perceived impairments in our children. It's unclear whether the deliberate avoidance of a perceived defect counts as an "enhancement," although it can be characterized as an attempt to shape the nature of our children. Whatever label we place on it, however, these techniques remain very blunt instruments.

Thus, at present the ability to pick and choose a variety of specific traits and capacities, to "engineer" oneself or one's children, remains more of a hope (or nightmare, depending on one's perspective) than a reality. Why, then, do I want to discuss what our policies should be regarding enhancements? First, the odds are that eventually we will develop a wide array of significant pharmacological enhancements. Already we have made some substantial progress, and although it may be another decade or three, pharmacological enhancements are likely to become available within the lifetime of your children, if not your own lifetime. (Genetic enhancements are less probable, at least in that time period, for the reasons indicated.) Among other reasons for this prediction is the fact that there is a drive to develop therapies that also will likely have the potential to be used as enhancements. For example, major companies and universities are expending significant resources on finding therapies that will ameliorate the symptoms of, if not cure, Alzheimer's disease. A drug that will assist persons who are losing their memory or other cognitive abilities also will likely improve the memory and cognitive abilities of those possessing capacities in the normal range. Similarly, drugs that will restore muscle function may also improve the muscle function of those not suffering from a disease or injury. As indicated, drugs developed for therapeutic purposes, such as Prozac and Ritalin, are already being used by persons without any clinically confirmed emotional or mental deficit.

This leads to the next reason for discussing our policies on enhancements. Opponents of enhancements—from technophobe extremists such as Jeremy Rifkin to respected scholars such as Sandel—have already begun deploying their arguments. Severe restrictions or bans on enhancements have been proposed. The Council of Europe has adopted a treaty that bans human germline modifications and human genetic enhancements in countries that sign and ratify the treaty.[13] Organizations in the United States, including various religious bodies, have called upon Congress to adopt similar legislation.[14] Development of enhancements could be slowed or stopped. For those who believe enhancements might be a good thing, this would be a highly regrettable occurrence. Moreover, because restricting or banning enhancements likely would also retard progress in developing some therapies, even those who are neutral about the desirability of enhancements have a lot at stake in this controversy. Accordingly, it is appropriate to consider what polices we should have on enhancements.

Before turning to our policy analysis, I have one final point. The line between therapy and enhancement can be difficult to draw, as the examples of Prozac and Ritalin show. Some question whether a line can be drawn, even in principle, whereas other have expended considerable intellectual resources in attempting to draw such a line.[15] Perhaps the most hotly debated issue in this area at the moment is whether aging is a "disease," which would make anti-aging drugs or genetic modification a therapy rather than an enhancement. Fortunately, for my purposes, whether a line can be drawn and, if so, what the criteria for the distinction between therapy and enhancement should be are not critical questions. If I were discussing health insurance questions, or arguing that genetic therapies but not enhancements are permissible, it would be difficult to avoid coming to grips with this problem, but I'm not.

For what it is worth, as with other questions discussed in this book, I believe the fact that it is difficult to know where to draw the line in all cases does not imply that a line does not exist. I believe there is a distinction between therapy and enhancement because, among other reasons, there are clear cases on either side of the line. One reason some doubt there is a distinction between therapy and enhancement is that our notions of what counts as a disease or a disorder are not always independent of our cultural norms. As Tristram Engelhardt points out in his charming and illuminating history of masturbation, masturbation was at one time regarded as a serious disease that could result in the death of the patient.[16] Until the 1970s, homosexuality was classified as a psychological disorder. Indisputably, cultural values can influence what is regarded as a disease. On the other hand, some conditions are diseases or disorders across all cultures. If I have uncontrollable diarrhea and vomiting and I am too weak to move from my bed, I have a disease of some sort, whatever culture I may be living in, and an antibiotic that would stop the diarrhea and vomiting and allow me to leave my bed counts as a therapy, not an enhancement. Similarly, Lesch-Nyhan syndrome, a genetic disorder that results in uncontrollable self-mutilation and usually a premature death, is presumably a disorder in all cultures, and a genetic intervention that would cure this disorder counts as a therapy.[17]

Therapies and enhancements do share a common thread that plays a role in my argument, and that is the fact that both are means of intervening in the world in an attempt to improve our situation and exercise more control over our lives. But more on that momentarily.

MORAL AND FACTUAL BACKGROUND

As with other topics, before evaluating arguments for and against pharmacological and genetic enhancements, I will set forth moral and factual premises that are accepted by the vast majority of persons in the contemporary developed world.

It is morally permissible to take measures to improve our living conditions in the sense of shaping our environment to protect ourselves against a range of hazards, such as cold, heat, floods, storms, drought, lack of nutrition, disease, injury, and so on. Certainly, this seems to have been an accepted premise throughout the existence of our species. Indeed, its acceptability is such that most humans probably have never given much thought to the premise itself, whereas they have given much thought to ways of safeguarding ourselves against hazards.

It is also morally permissible—all other things being equal—to take measures, including the development and application of new technologies, that facilitate our accomplishment of tasks deemed necessary or desirable. Again, this premise seems incontestable. From the time when humans began keeping records (and before, to the extent archeological evidence permits inferences), humans have devised new, improved means of achieving their objectives. We have made the task of moving from point A to point B safer, more reliable, and tremendously more rapid—recent experience with commercial air travel notwithstanding. As with transportation, so too with communication, from the invention of writing to the printing press, telegraph, telephone, e-mail, and cell phones. Much the same can be said for other areas of human endeavor, such as agriculture, construction, production of energy, or healthcare. Obviously, new technologies can have adverse effects and be put to immoral uses. The use of cell phones to set off bombs remotely is just one example. But we can evaluate new technologies as they become available and make a reasonable calculation of costs versus benefits and determine whether the new technology should be banned or heavily regulated. On balance, their occasional use by terrorists does not indicate we should not allow cell phones to be readily purchased. On the other hand, allowing Wal-Mart to sell uranium enrichment devices is probably not a good idea.

Shaping our environment and developing new technologies—improving the way in which we live—are presumptively morally permissible. How about directly improving ourselves or our children through the acquisition of new traits and abilities? Is this permissible? Here I am going to avoid the easy answer of simply saying "yes." It is true that in contemporary democracies, self-improvement seems permissible if not obligatory. Much the same can be said for improving our children's abilities. Think of the billions of dollars spent each year on education, much of it from public funds. And the funding is not limited to supporting the acquisition of knowledge. We also expect our schools to improve our children's athletic and social skills. Moreover, for those parents unsatisfied with what the public school system offers, there is the alternative of private education. Some regard private schools as the preserve of the wealthy. Not so. Although the median income of families with children in private schools is significantly higher than the median income of families with children in public schools, some families make tremendous financial sacrifices to enable their children to have the perceived benefits of a private education. Typically, we regard such sacrifices as praiseworthy.

Of course, adults use education to improve themselves as well as their children. We enroll in a myriad of degree and nondegree courses and programs. Moreover, although education is the most obvious example of self-improvement, other examples come readily to mind, such as the efforts individuals make to improve their character, temperament, habits, and general situation in life.

However, it is important to point out that self-improvement or the improvement of one's children has not always been regarded as acceptable for all people. To the contrary, throughout history many societies have adhered to norms that have kept people in their "proper place." Slaves, serfs, and members of lower castes not only were not encouraged to educate themselves, but also were often punished for trying to do so. And, of course, throughout most of human history, self-improvement for most women was confined to very narrow areas, such as cultivation of social graces or acquisition of skills such as weaving. These examples are familiar ones, but it may surprise some readers to learn that as recently as the late 1800s in countries as liberal as England, certain attempts at self-improvement were frowned upon, if not openly discouraged. For example, Thomas Hardy, the famous English novelist, came from a lower-class family that did not have much in the way of formal education. When his parents apprenticed Hardy to an architect, at significant cost to themselves in the hopes of having their son move up the social ladder, the reaction of the local vicar was to preach a sermon denouncing "the presumption of members of the lower classes who aspired to join the professions."[18] Rigid class distinctions, which limited what persons could do to better their lot in life, prevailed in many countries until well into the twentieth century.

One reason I bring up this history is that it should be borne in mind when we discuss in more detail objections to enhancements. As indicated, some object to enhancements because they supposedly manifest a desire to control or shape our destiny. Others maintain that they will overturn social arrangements currently accepted as equitable. Not often noted is that by allowing persons the opportunity to improve their traits and capacities rapidly, enhancements may disturb a status quo that some opponents of enhancements find comfortable and desirable. Perhaps part of the opposition to enhancements may be traced to the sentiment that it is "presumptuous" of a person of average intelligence to enhance her cognitive abilities to the level of those scholars who author rhetorically impressive books criticizing enhancements. Those of average ability should be satisfied with their lot and not aspire to more.

Hold that thought. For now, I take it as a given that in contemporary developed countries self-improvement or the improvement of one's children through means such as education is morally permissible. The issue is whether improvement through means such as pharmacological or genetic enhancements is somehow impermissible, and, if so, on what basis are such means of self-improvement distinguishable from accepted means.

WHAT IS WRONG WITH TRYING TO ALTER THE WORLD AND IMPROVE ONE'S LIFE?

We have seen that Sandel has argued against enhancements because they constitute an attempt to "master" nature. For him, enhancements "represent a kind of hyper-agency, a Promethean aspiration to remake nature, including human nature to serve our purposes and satisfy our desires."[19] If this is the crux of Sandel's objection—and I believe this is a fair reading of his arguments—then his objection is a few millennia too late. His reference to Prometheus unintentionally reveals the fundamental flaw in his objection. Human history *is* the history of our efforts to remake and shape nature. Prometheus, of course, was the mythical Titan who gave fire to humans and taught them how to control it, along with other skills. The Olympian gods were angry with Prometheus because humans were no longer helpless and utterly dependent on them.

The story of Prometheus expresses a basic truth about human existence, and that is that most humans have sought to control and shape their environment in order to achieve their objectives. Furthermore, unlike Sandel, most humans see nothing inherently immoral about this desire. If early humans had listened to the Sandels in their midst, we'd still be naked, exposed to the elements, and living off whatever carrion, leaves, and berries were within our reach.

This is an appropriate juncture at which to note that Sandel frames his debate over enhancements improperly. At times he suggests that the desire for enhancements is the desire for "mastery." As he states, "The problem [with enhancements] is . . . the drive to mastery. And what the drive to mastery misses, and may even destroy, is an appreciation of the gifted character of human powers and achievements."[20] The suggestion is that our attempts to achieve some measure of control over nature are motivated by the desire to stand on a mountaintop, thump our chests, and say, "We did it." But few seek control over nature for its own sake. We do not seek the status of "masters." We seek to shape our environment as a means to an end. We want shelter and electricity not so we can stare at the light and feel proud, but so we can read Sandel's book in the dark—presumably a goal that meets with his approval.

Just as he mischaracterizes the desire for shaping our environment, Sandel similarly mischaracterizes the results of using enhancements. He suggests that it is important to retain the sense that there are some matters that remain outside of our control. To use his terminology, we must retain the sense of "giftedness" or the sense that our point of "origin . . . exceeds our control."[21] Supposedly, this sense of giftedness will be lost if we use enhancements to improve our traits and capacities.

However, assuming that a sense of giftedness is important, perhaps as a deterrent to overweening pride, no reasonable person will lose that sense. Every individual's point of origin necessarily exceeds her control. We are the products of billions of years of biological evolution and millions of years of hominid evolution. Moreover, our ability to develop pharmacological and genetic enhancements is possible only because we stand on the shoulders of countless generations of humans who strove to control their environment and improve themselves and their children. Finally, one thing that enhancements cannot possibly change: we will never be able to choose our parents and we will never have control over the genetic endowment they provide, whether by choice or chance. Thus, a radical contingency pervades and will continue to pervade every human's existence. To the extent it is important to realize we are the results of events we did not control—that we are the products of fortune or "gifts"—enhancements will not affect this understanding. For some unexplained reason, Sandel mistakenly believes that enhancements will obscure this reality, but that would be true only if enhancements led to delusions—in which case they hardly would qualify as enhancements.

At bottom, the principal claims of Sandel's argument against enhancements are merely a variant of the tried-and-false appeal to nature. The natural is good; the unnatural is bad. By this point in the book the reader should be able to discern why such an appeal to nature is as uninformative as it is unpersuasive. At best, it is metaphor masquerading as reasoned argument.

Some may think that I am unfair to Sandel because the primary motivation for his book appears to be his opposition to genetic or pharmacological enhancements, not our accepted ways of "mastering" nature, such as by constructing shelter or finding a faster way to get somewhere. But that is not how he formulates his argument, which is as an objection to the drive to "mastery" in general. In fact at one point he states, "Sometimes mechanical enhancements [such as sound amplification systems] can be more corrupting than pharmacological ones."[22] Sandel's misguided focus on the supposedly corrupting influence of the Promethean drive for mastery, if extended to its logical conclusion, would prohibit medical therapies as well as enhancements, because therapies attempt to overcome nature as much as enhancements. Recall a while back I mentioned that although I believe therapies can be distinguished from enhancements, they have something in common. What they have in common is they both represent efforts to shape and improve our lives by manipulating the world for our benefit. Both therapies and enhancements do not accept the world "as is." Instead, they refashion the world to remove barriers to our flourishing. Even so-called "natural" remedies, such as herbs, are the result of human intervention in the world and exploitation of other organisms for our benefit. If we truly renounced the Promethean desire to remake nature, then we should rely exclusively on our immune systems and our bodies' inborn repair mechanisms to restore ourselves to health. Somehow I do not think this is a program Sandel will be able to sell to most people.

Granted, Sandel does state that technology that brings out our natural gifts is permissible, whereas technology that "distorts and overrides natural gifts" is troubling;[23] however, he does not provide any meaningful criteria for distinguishing between the two sorts of technology. Is scuba diving equipment technology that brings out our natural gifts or technology that distorts them? What about designing a human with gills that achieve the same result as the equipment? Or, to take a medical example, what about gastroplasty (stomach stapling)? It is difficult to see how such a procedure brings out any natural gift, but studies have shown that it is effective in reducing the mortality of patients from obesity-related diseases.[24]

Nonetheless, it is apparent that Sandel does think many therapies are permissible. At one point he expressly states, "The moral quandary arises when people use . . . therapy not to cure a disease but to reach beyond health."[25] So even if his position lacks coherence, we should examine the claim that there is something about enhancements other than the fact that they are the product of a desire to master nature that makes them morally questionable. Let's consider the objection that enhancements are bad because they change our biological nature and undermine human agency. This is how some leading opponents of enhancements, such as Leon Kass, characterize the issue.

In his argument against enhancements, Kass contends that genetic and pharmacological enhancements can be distinguished from education as a means of self-improvement because "biomedical interventions act directly on the human body and mind to bring about their effect on a subject who is not merely passive but plays no role at all. He can at best *feel* their effects *without understanding their meaning in human terms*."[26] He goes on to state that with biomedical enhancements, "The relations between the knowing subject and his activities, and between his activities and their fulfillments and pleasures, are disrupted."[27]

In response, I would say that Kass both misstates how accepted, conventional means of self-improvement work and how genetic and pharmacological enhancements would work. First, education and training are themselves biological processes. A teacher, a tape that you listen to, a DVD that you watch, all aim at bringing about changes in your brain, in establishing new neurological pathways that ultimately will allow you to perform tasks you would have been unable to perform prior to the education and training. Of course, other parts of your body, such as your muscles, may be affected as well. But there is no mystical, nonbiological process involved. Knowledge is being injected into your brain; it is simply being injected through sensory inputs instead of a drug.[28]

Furthermore, we are passive and we may have no understanding of the point of our education when we first begin to learn. This is clearly true when we are very young, but even when we are older we may be passive and unable to see the point of the training until we become better acquainted with the skill in question. Have you ever tried to teach chess or baseball to an adult who has never engaged in or even observed either activity? You will find that at least initially in the pupil's mind there is little perceived relationship between the training and the end sought.

Turning to enhancements, Kass exaggerates and misrepresents the effects of such interventions. A drug that will allow you to read and understand a textbook on calculus in a day's time would be an incredible breakthrough—something clearly beyond the power of any drug available in the foreseeable future. Even such a powerful drug, however, will not destroy the connections between "the knowing subject and his activities, and between his activities and their fulfillments." The person still has to apply his knowledge and utilize the skills he has acquired so rapidly. Drawing a good hand in poker does not guarantee victory; one still has to play the hand one has been dealt. Similarly, one has to make appropriate use of one's enhancements.

Enhancements will not eliminate achievements; they will make them more accessible and allow persons to achieve more in less time than their predecessors. Conceivably, persons could acquire the knowledge of a PhD in physics in a few weeks time. But would that be such a bad thing? Perhaps the survivors of earlier generations will lament that they did not have all the advantages of Generation E (for enhancement), but in this respect they will be no different from the generations that have gone before them. For the most part, later generations have been able to do more and do it quicker than earlier generations (although the Middle Ages were a period of retrogression in the West). Compare your situation as you sit in your centrally heated, air-conditioned home, with its computer, state-of-the-art kitchen, and so on, with the situation of a human in Neolithic times. You can do a lot more things much more rapidly than that individual. Our cave-dwelling ancestors could not even imagine or comprehend the range of our activities. To the extent they could, they might, along with Kass, assert that we cannot experience these activities as "our own as human."[29] After all, we passively rely on the computer, the air-conditioning system, and so on to do so much for us. But that would be to mistake the means for accomplishing the goal with the goal itself.

What Kass does not understand is that changing the means by which we accomplish a goal does not eliminate the goal. Much less does it eliminate us as human agents.

Furthermore, our currently accepted technology is not limited to devices that allow us to engage in the same sorts of activities as our ancestors, only more effectively and rapidly. We have developed devices that directly extend our sensory capacities. Think of x-rays, CT scans, night vision goggles, sonar, radar, and so forth. All of these technologies allow us to "see" or "hear" objects and events that were imperceptible to our ancestors. No one seems to object to these technologies. Why should we object to pharmacological or genetic enhancements that provide us with similar capabilities? What exactly is the moral difference between wearing a pair of night vision goggles and taking a drug that provides one with similar powers of perception? I fail to detect any.

Enhancements do have the potential for transforming our world. Moreover, we must admit that, if enhancements are as robust and efficacious as some predict, the extent of that transformation can be only dimly perceived. This, I believe, is at the root of the concern some have over enhancements. They will change our way of life. But if enlarging our ability to shape and give direction to our lives and accomplish our objectives is a good thing, they likely will change our way of life for the better.

This may seem like faint reassurance given the anxiety that uncertainty usually causes. However, we do have a precedent to rely upon. For a few decades now, we have already had available a pharmacological enhancement that has caused a revolution in our social and economic relations and transformed the potential of half of humanity. Although many decried the enhancement when it was first introduced, and there are still some who reject it as "unnatural," nowadays most hardly give it a second thought and regard its availability as a given. I am referring to oral contraceptives.

Some may question whether oral contraceptives qualify as an enhancement. Of course they do. They are drugs that provide women with the power to prevent pregnancy. Moreover, they provide this power very quickly. If this does not qualify as an enhancement, what does?

Granted, for some women they have a therapeutic effect as well (they can be used to treat endometriosis and acne, and obviously they help preserve the health of women physically incapable of bearing a child), but these are secondary to the effect they have as an enhancement. Moreover, as we have discussed, many projected enhancements (memory pills, etc.) will have therapeutic effects as well. The status of oral contraceptives as an enhancement may be obscured by the fact that most health plans provide coverage for them, just as they do for prescribed therapeutic drugs, but this reflects the market power of women more than medical fact. Undoubtedly some projected enhancements, provided they can be produced relatively inexpensively, will be covered by health plans as well.

Not only are oral contraceptives an enhancement, but they are probably the most consequential enhancement we will witness for some time. Having a reliable means of preventing pregnancy vastly increased the ability of women to control the direction of their lives and enabled them to accomplish much more than they would have been able to otherwise. It's not just a matter of sexual freedom. Look at law offices, physicians' offices, and corporate boardrooms today and compare them with similar establishments in the 1950s. Some will attribute the near total absence of women in professional positions prior to the 1960s to sex discrimination, but that is only a small part of the explanation. The attitude that women were not suited to such positions was based partly on the fact that few women pursued such positions, and relatively few women pursued such positions because they could not reliably control whether and when to have children. With birth control pills, women no longer had to choose between having a career and a relationship.

Women were the direct beneficiaries of enhanced control over their bodies that oral contraceptives provided, but society as a whole also has benefited greatly from this enhancement. The entry of women into careers previously closed to them has resulted in significant contributions of intellectual capital and has provided a great boost to our economic productivity. There were some social stains caused by the massive influx of women into the workforce, especially into professional positions, but overall we have adjusted well to these changes. Having to devote judicial and administrative resources to sexual harassment complaints and lawsuits—which were occasioned by the mixing of men and women in employment settings—seems a small price to pay for the benefits the Pill has made possible.

The case of oral contraceptives shows we can have an enhancement that brings about far-reaching social and economic consequences, as well as inducing changes in the behavior of millions, and not only survive the experience but welcome its effects.

In sum, enhancements promise to improve our abilities or traits, or the abilities or traits of our children, much quicker and more reliably than other methods of self-improvement. This hardly seems to provide a sufficient justification for banning them. But the literature on enhancements is vast, and some may question whether I have confronted the strongest objections. I will try to answer briefly some other objections in the next section.

A QUICK REVIEW OF SOME ADDITIONAL OBJECTIONS TO ENHANCEMENTS

Oral contraceptives are instructive because this historical example helps illustrate why some of the common objections to enhancements lack any merit. Let's replay Sandel's and Kass's principal objections first. Oral contraceptives do not bring out any natural gifts; to the contrary, they suppress ovulation. So what? This does not show they are immoral. They act directly and quickly on the body, and the role the woman plays in avoiding pregnancy is arguably diminished; the drug substitutes for abstinence and other pregnancy-avoiding behavior. Again, so what?

Some have argued that enhancements will destroy character because they will allow persons to obtain their goals too easily. This objection is implicit in Kass's argument and it is explicit in Francis Fukuyama's criticism of enhancements that improve a person's confidence and self-esteem (such as Prozac when used for nontherapeutic purposes). Fukuyama argues that such enhancements shortcut the "normal, and morally acceptable, way of overcoming low self-esteem," which is "to struggle with oneself and with others, to work hard, [and] to endure sometimes painful sacrifices."[30] Something similar could be (and was) said for oral contraceptives. Oral contraceptives not only threatened to produce libertines, but were a way of shirking the challenge of being both a mother and a professional. Birth control is for slackers. In both cases, the criticism is unpersuasive. Not having to deal with one obstacle to self-fulfillment does not remove all challenges; it just allows one to concentrate on the challenges that one wants to face, as opposed to being distracted and weighed down by pointless hindrances. Having a "room of one's own," to quote Virginia Woolf, may give a woman more opportunity to write. It does not relieve a woman of the burden of writing. Whatever benefits an enhancement may provide, there will still be goals that the enhanced person must achieve.

A related objection is that enhancements will cheapen the accomplishment, even if they do not affect the character of the person taking the enhancements. This is a common objection put forward by the opponents of enhancements. Almost invariably, the point is illustrated by an allusion to athletic competitions.[31] It is true, as the allegations of Barry Bonds's steroid use or the multiple scandals involving cyclists in the Tour de France demonstrate, that many dismiss the accomplishments of athletes who are believed to have cheated in some way. But life is not a game. Athletic competitions are a carefully circumscribed subset of human endeavors with specific, detailed rules that define what constitutes permissible conduct. These rules are not, of course, limited to prohibiting ingestion of substances that provide an advantage, but they encompass any technique not generally available to one's opponents or that reduces the element of chance in the competition, such as videotaping an opponent's signals.[32] The bodies in charge of athletic competitions impose these rules for obvious reasons: principally, to keep the competition interesting to the spectators and fans. No one is going to pay five hundred dollars a ticket to see the New England Patriots take on a group of kids from kindergarten. Regulating enhancements may pose a special problem for athletics, but that does not imply they should be banned altogether. Suppose someone was able to devise a cheap, safe means of obtaining energy through fusion because they took drugs that enhanced their cognitive skills. Would we be outraged—or overjoyed that our energy crisis has been resolved? Currently, many individuals owe their accomplishments to factors such as a better education, financial independence that allows them to concentrate on a particular goal, or the availability of a battalion of supporting researchers. We do not strip scientists of their Nobel prizes merely because they were better educated and had more time and resources at their disposal than other scientists. And, again, the example of oral contraceptives confirms that accomplishments are not necessarily cheapened by enhancements. If a lawyer made partner in a major law firm in part because she was able to postpone her child-bearing years until her late thirties, does this entail she did not "deserve" to become a partner?

Some have argued that the availability of enhancements will place pressure on some to use enhancements even though they would prefer not to use enhancements.[33] That may be true in some cases, but this hardly seems sufficient to ban enhancements. Which is the greater restriction on individual autonomy: the indirect pressure someone may feel to conform to widely accepted patterns of conduct, or prohibiting everyone from engaging in the conduct in question? We encounter indirect pressure all the time to conform our conduct to prevailing patterns, whether it is a question of continuing our education, wearing the latest fashions, voting for certain political candidates, avoiding "offensive" behavior, and so on. We may endorse this pressure in some cases and condemn it in others, according to our preferences. What we do not generally do is prohibit persons from engaging in conduct X merely because others may feel peer pressure to follow their example. The fact that some would prefer to do without oral contraceptives but feel pressured to take them because that is what is expected in their social circles clearly does not warrant a ban on oral contraceptives.

And then we have what is effectively the flip side of the previous objection. Some have argued that enhancements may be self-defeating because if everyone uses an enhancement, then whatever competitive advantage the enhanced individual wanted will be lost.[34] However, this assumes that what persons want out of enhancements is a competitive advantage. This assumption is false, at least in many cases. Although being able to use an enhancement such as oral contraceptives assists women in pursuing a career, most women take oral contraceptives so they can exercise some control over their own lives, not to gain an advantage over women who do not use birth control. Similarly, a person may want to enhance her intelligence to understand calculus, improve her chances of finding a cure for cancer, obtain the ability to repair her own computer, or acquire the means of penetrating George W. Bush's tangled syntax, without any thought to whether this will enable her and her progeny to acquire more wealth, prestige, and power than others.

Of course, enhancements may provide competitive advantages whether they are desired for that reason or not. The next section of this chapter will discuss what restrictions, if any, should be placed on enhancements because of their societal impact.

Before closing this section—which is devoted to assessing whether there is anything immoral about enhancements regardless of their effect on society—I do want to discuss enhancements for children, which has received particular attention from opponents of enhancements. The dust jacket for Sandel's book even features pictures of babies.

Is there something special about choosing enhancements for our children as opposed to choosing enhancements for ourselves? Well, the obvious difference is our children have no say in the matter. Opponents of enhancements protest the advent of "designer" babies and contend that by selecting enhancements for our children we are depriving them of their freedom. Environmentalist and technoskeptic Bill McKibben has proclaimed that genetically engineered babies will be left without any choices at all. Parents who use enhancements will have decided "for once and for all, certain things about [their baby]: he'll have genes expressing proteins that send extra dopamine to his brain to alter his mood; he'll have genes expressing proteins to boost his memory, to shape his stature."[35]

In rebuttal, a couple of preliminary points. As is true with many persons writing in this area (both pro- and anti-enhancement), McKibben seems to subscribe to a genetic determinism. In other words, genes determine everything about us. That's wrong. Our genetic composition does greatly affect our traits and abilities, but how those traits and abilities are manifested depends on a person's environment. Also, for reasons already discussed, genetic enhancements are much less likely than pharmacological enhancements. But opponents of enhancements typically focus on genetic enhancements because they strike a more responsive chord. For many, messing with someone's DNA seems more disquieting than offering a drug.

But let's turn to substance. McKibben and others draw a false contrast between the way the characteristics of our children are determined now and how they will be determined in a future with enhancements available. The choices we make for our children without the benefit of enhancements already determine much about them. We decide whether to nourish them from the breast or bottle, whether and where to place them in daycare, whether to send them to piano or swimming lessons or both, whether to indoctrinate them in a particular religious faith, and so forth. Many of these choices have profound and long-lasting consequences. Whether to provide enhancements represents simply one more set of choices we will make for them. Furthermore—and this is a key point all too often ignored—biological parents *already* determine the biological makeup of their child. The fact that they may not know what genes they are providing does not remove the reality that they jointly determine the genetic composition of their child. Finally, contrary to the common assumption that enhancements will influence the makeup of our children far more than conventional choices, this is not necessarily true. Enhancements may well be reversible in a way that conventional training and education are not. For example, once we have an understanding of how certain substances affect one's cognitive abilities, drugs can be used to increase or decrease one's cognitive skills. The blue pill raises your IQ by ten points; the red pill lowers it by ten points.*

Admittedly, enhancements will make your child's makeup more pre-

*Any correlation between this example and so-called blue states and red states is purely coincidental.

dictable. Sandel and others who oppose enhancements often argue that the ability consciously to shape our children's traits, character, and capacities more than we do now threatens to undermine the unconditional love we have for our children.[36] First, I question whether our love for our children is always unconditional. It is undoubtedly true that the emotional bond most parents have with their children is very strong and most parents will both sacrifice much for the sake of their children and forgive them much. But love can be tested and lost, sometimes for good reasons, sometimes for bad. The Menendez parents, had they survived the shotgun blasts their sons delivered to their heads, understandably may have lost some of their affection for their children.[37]

Second, and more important, children repeatedly fail to live up to their parents' expectations now, as assuredly as they have throughout our existence as a species, without placing in jeopardy their parents' love for them. Why would it be any different if the parents were able to select a few traits? If anything, if the enhancement failed to take hold in the desired way, the parents would blame themselves, *not* the child.

But, of course, that leads to the other concern. (Opponents of enhancements often try to have it both ways.) Sandel asserts that parents may feel more responsible for the makeup of their children if they have more control over their characteristics.[38] Is this feeling of responsibility a bad thing? Fifty or sixty years ago it was not understood how smoking or drinking during a pregnancy might affect a child. Now we know, and mothers who are responsible take the appropriate precautions. Which person would you rather be: the carefree mother who in 1950 did not realize her cigarette and gin might adversely affect her child or the mother who today, tempted though she may be, scrupulously avoids tobacco and alcohol because of what Sandel characterizes as the "burden" of responsibility? To me, there does not seem to be much contest.

Finally, Sandel and others who oppose enhancements state that it is a blessing to regard children as the creations of "nature, God, or fortune" and enhancements will eliminate that blessing.[39] This is more poetry than argument, but let's try our best to analyze it. First, if there is a deity, no one can say what that deity's position is on use of enhancements, and it is unclear why using the skills the deity has made available to us to enhance our children's capabilities make them any less creatures of "God." (See the discussion in chapter 1.) And nature? Whether children are generated without a thought to enhancements or are enhanced, they will be products of nature. You can't get more natural than genes, and chemical enhancements will adhere to the laws of nature just like everything else. I am unaware of any technology that can deliver a supernatural enhancement.

This leaves us with fortune. If by saying that at present children are creatures of fortune, Sandel means to say that parents do not cause their children to have whatever abilities and traits they end up with, this is largely false. All of us are the products of our genes and our environment. As I have already pointed out, our biological parents contribute their genes and (most of the time) shape our early environment. If by fortune Sandel means that parents do not know how their children will turn out, that is true. It is also accurate to assert that enhancements will reduce some of that uncertainty. So the contrast is not between having a role in producing a child with various characteristics and having no role in producing a child with various characteristics. The contrast is between producing a child based at least in part on informed choices and producing a child by the equivalent of spinning a roulette wheel while blindfolded. When the contrast is phrased this way, I'm doubtful whether it is a "blessing" to subject our children to the whims of fortune. If you had the option of spinning the roulette wheel or choosing a set of enhancements that more likely than not would result in your child possessing the cognitive skills to undertake complex tasks, find fulfillment in many different intellectual pursuits, and prosper in most professions, which choice would you make? Gambling with your child's life does not seem particularly prudent, responsible, or virtuous.

The failure to recognize that we are already determining our children's biological makeup, albeit haphazardly, and that unpredictable and random mutations have affected the development of our species, has resulted in some breathtakingly benighted proposals from some bioethicists. As I have indicated, unfounded concerns over "designed" children have already resulted in a ban on inheritable genetic modifications in some countries. In an extraordinarily overwrought essay, George Annas, Lori Andrews, and Rosario Isasi have called for an international treaty banning all germline modifications (that is, both enhancements *and therapies*). They claim that "inheritable genetic alterations can be seen as crimes against humanity of a unique sort: they are techniques that can alter the essence of humanity itself . . . by taking human evolution into our own hands."[40] Think about that claim for a moment. First, would it be a "crime" against humanity if we were able to develop germline modifications that provided immunity against various diseases and also ensured that mutations that produce horrifying disabilities would not occur? If such conduct be a crime, enroll me with the felons.

Second, reproducing organisms, including hominids, have always taken evolution into their own hands; they just may not have been able to predict the outcome. Was it a "crime" for Australopithecus to evolve? Was it a "crime" for Cro-Magnon to supersede Neanderthal? Genetic composition of any set of organisms, including ourselves, is not static. It is always changing and will continue to change, regardless of any treaty. The only way to preserve humanity as is would be to stop having children, whether designed or undesigned, and that seems like an extreme measure. The issue is not whether we alter our genes; the issue is whether we alter our genes at random or use our intelligence to promote changes that will be beneficial. I don't think Annas, Andrews, and Isasi have made a persuasive case that random mutations, the equivalent of throwing a coin into a wishing well, are so advantageous over intelligently guided genetic alterations that the latter constitute a "crime."

In conclusion, there is no morally cogent basis for severely restricting or prohibiting pharmacological or genetic enhancements (or, for that matter, germline modifications that are therapeutic), assuming they can be implemented safely. The rejection of enhancements stems principally not from sound moral reasoning, but from the instinctive aversion that many feel to this new technology. At one point in his essay, Kass candidly admits that it is difficult for him to put his "disquiet into words." He continues, "We are in an area where initial repugnances are hard to translate into sound moral arguments."[41] I contend that is because there are no sound moral arguments against enhancements. At least I am unaware of any moral theory that forbids individuals from improving themselves by enhancing their capacities to engage in worthwhile activities, absent direct, imminent, and serious harm to others resulting from their efforts at self-improvement. Unlike Thomas Hardy's vicar, I do not believe it is presumptuous for persons to better themselves.

ENHANCEMENTS AND JUSTICE

However, as previously indicated, there are those who want to ban or impose onerous restrictions on enhancements not because they are impermissible by themselves, but because of the harmful effects they are predicted to have on society. Much ink has been spilled by scholars devising social arrangements to counter what they perceive as the threat posed by enhancements to a just society. The concern most often cited is that enhancements will be distributed unfairly, and there will be huge gaps in income, wealth, political power, and other social goods between those who are enhanced and those who are not. To prevent this from happening, these scholars urge us to formulate or even implement regulations on enhancements now. Uncertainties about how enhancements will actually affect society are brushed aside. As one group of scholars has observed:

> In contemplating the disturbing challenges that the possibilities of genetic intervention pose for our traditional ways of thinking about justice, it is tempting to conclude that we are ill equipped to make any firm judgments about what justice requires. This temptation, however, ought to be resisted. Some conclusions can indeed be drawn about the requirements of justice in the genetic age.[42]

To the contrary, I urge us to yield to "this temptation." The fact of the matter is we are ill equipped to make firm judgments about the requirements of justice in a society transformed by enhancements. Those who hold otherwise do not sufficiently take into account the probability that as we transform ourselves through enhancements, our notions of justice also will be transformed.

One does not have to be a relativist in ethics to accept that conclusions about what social arrangements we should have, and how we should distribute wealth, political power, and other social goods, may be affected by the conditions of the society in which we find ourselves. Right now, would you agree to an arrangement in which you were obligated to stay in one location, work the land, and provide part of the fruits of your labor to someone who lived in much more comfortable conditions? Presumably not. But if you lived in early medieval Europe and the person to whom you provided your produce also used the weapons and armed men he had at his disposal to guarantee your safety from marauding, merciless bands, you might not find this arrangement unfair—especially when the alternative was to live on your own, with no protection from a central government, and meet an early death. No sane person would suggest a return to feudalism under our current conditions, but to criticize feudalism as "unfair" is anachronistic.

Terms such as "just" or "fair," when used to describe how we divvy up social goods, cannot possess any meaningful content outside of a given framework. Those who prescribe regulations for some future society in which significant enhancements become available assume that these enhancements will have a very substantial impact on our social and personal relationships. But if that is true, we cannot now determine what principles of distributive justice should apply in such a radically transformed society.

One might wonder what the harm is in speculating about how we might regulate enhancements. There is no harm in such speculation, by itself. But those who propose regulations often want them to be taken seriously. As I have indicated, there are already treaties in Europe that ban genetic enhancements (in signatory countries) and some opponents of enhancements have pushed for legislation or regulation in the United States. Restrictions on enhancements based on concerns about distributive justice are premature at best. At worst, they will impede the development of enhancements that could provide important benefits to many, if not all, of us.

Moreover, although speculation about the requirements of justice is not in itself harmful, it may divert attention from more immediate and, one would think, more obvious problems of distributive justice. In this regard, it is paradoxical that those who favor strict control of enhancements tend to emphasize the disproportionate advantages likely to accrue to the wealthy in the absence of regulation or prohibition. If future hypothetical disparities between the wealthy and the rest of us are a cause for concern, one would think that present actual disparities should be of as much, if not more, concern. There is no need to await the dawning of the enhancement millennium to take action; it would be far simpler and less morally problematic to redistribute wealth now than to distribute enhancements—especially genetic enhancements—later. In sum, the requirements of justice in an enhanced future can be perceived only dimly, and squinting at this distant horizon not only may be a futile exercise, but also a distraction from problems that confront us today.

Perhaps the best illustration of the difficulties of devising regulations that would ensure a just society in an era of enhancements can found in the work of bioethicist Maxwell Mehlman. I pick Mehlman not because his suggestions are the worst of the lot. To the contrary, they show much thought and careful consideration. The problems created by his proposed regulatory scheme underscore the difficulties of devising rules and principles for a society not yet in existence even when these rules and principles are formulated by someone knowledgeable about law and ethics.[43]

I observed at the beginning of this chapter that Mehlman believes we should ban germline genetic enhancements. More on that position later. With respect to somatic genetic enhancements (that is, enhancements that affect only the individual enhanced as opposed to her descendants), Mehlman argues that we should permit such enhancements, *provided* certain conditions are met. Mehlman is concerned, as are many others, that enhancements will provide the enhanced with competitive advantages in virtually all areas of life. "The object of the competition may be any desirable good: money, employment, status, affection, sexual favors, political influence, or market power."[44] Accordingly, he maintains that we should condition the right to acquire somatic genetic enhancements on a commitment by the person seeking enhancement to employ her "abilities in some predetermined socially beneficial manner."[45] The government would grant enhancement licenses and punish persons who have illegal enhancements or fail to fulfill the terms of their license by depriving them of the enhancement or imposing "various forms of social handicapping, surtaxes or monetary penalties, and . . . imprisonment."[46] Mehlman believes there is some precedent for his proposal because his enhancement "licensing system would be similar to legally enforced professional licensing schemes that give their holders powers and privileges denied ordinary citizens in return for agreements to abide by rules designed to promote social goals and to refrain from behaving in socially undesirable ways."[47] Mehlman's latest work on this topic confirms that he intends to use fiduciary relationships, such as those between physician and patient and attorney and client, as a model for his enhancement licensing scheme.[48]

Mehlman's proposal is very creative, but it presents serious practical and moral problems. To begin, his attempt to analogize his enhancement licensing scheme to the regulations imposed on professionals, such as lawyers or physicians, just doesn't work. Without in any way being critical of lawyers or physicians as groups,[49] the regulations imposed on them leave ample room to engage in conduct that many would regard as socially undesirable and do little by way of requiring them to promote social goals. For example, I am unaware of any lawyer ever being disbarred for refusing to provide free legal services to the indigent and for devoting her entire professional life to reducing the tax obligations of the wealthy and powerful. Similarly, physicians may, without penalty, devote their entire professional lives to providing cosmetic surgery for the wealthy and powerful. Moreover, lawyers and physicians are free to compete among themselves with minimal restrictions bordering on the inconsequential. Essentially, current regulatory schemes merely command persons with special authority to be careful about protecting the interests of their patients/clients and prohibit them from manipulating their patients/

clients for their own self-aggrandizement. Requiring those who would be granted enhancement licenses not to be grossly negligent when entrusted with certain tasks and prohibiting them from using their enhanced capacities to defraud others would not diminish significantly their ability to dominate the competition in their chosen fields and to acquire wealth, power, and prestige. In terms of its goals, the enhancement licensing scheme contemplated by Mehlman appears to go far beyond current regulations that condition one's right to engage in certain activities on one's compliance with guidelines that protect others only against the most egregious misconduct. Therefore, to the extent Mehlman is using current licensing schemes as models for his proposed enhancement licensing, I am afraid it will not accomplish what he hopes it will.

Of course, he is free to propose more stringent regulation, but the more stringent the regulation, the more difficult it will be to enforce and administer. Bear in mind that the proposed licensing scheme encompasses all manner of genetic enhancements and all possible competitive uses of enhancements, including enhancements used to triumph over rivals "for someone's affection."[50] To eliminate competitive advantages in *all* areas, one would have to monitor and control, for example, the writing, recording, and performance of music; the analysis, tracking, and investment of personal, as well as corporate, financial resources; the endurance, memory, wit, affability, and "charisma" of all political candidates; and the physical appearance and personality of all those seeking friends, lovers, and mates. The monitoring and control that the government would have to exercise over the lives of its citizens would be nothing short of Orwellian.

Moreover, wherein lies the moral justification for such pervasive control of one's capacities and activities? To my knowledge, there is no precedent in modern, democratic society for conditioning someone's liberty to seek self-improvement, funded with her own money, or that person making a significant commitment to use her new capacities in socially beneficial ways or to refrain from engaging in activities in which her capacities would provide her with a competitive advantage. As indicated, the social commitments required of lawyers and physicians are negligible.

The commitments required of those who engage in other activities, which may be no less important or competitive, are virtually nonexistent. I do not have to commit to providing free translation services because I have the means to take an intensive language course in Arabic, nor do I have to refrain from noting my abilities when I apply for a position with the Foreign Service. Similarly, I do not have to counsel others on how to have success with the opposite sex because I have had the benefit of both a liberal education and a hair transplant, nor do I have to scar myself or feign confusion over the meaning of a Woody Allen joke to "level the playing field" with others. Is there a sound moral argument to require the enhanced to obtain a license to romance?

In a response to an earlier article I wrote pointing out some of the difficulties in his proposed regulations, Mehlman admitted that there are a number of problems with his suggestions on regulating access to enhancements. Nonetheless, he emphasizes that he is not insisting that such restrictions on somatic enhancements be imposed now (they hardly could be since we do not have available the enhancements he proposes be regulated), but rather urging us to consider such restrictions before it is "too late." He contrasts his approach with what he characterizes as my "blasé attitude."[51]

Again, there is nothing inherently wrong about thinking about such issues. But drafting detailed plans—and Mehlman admits his proposals constitute plans, not just idle speculations—to deal with future situations that may never come to pass, and the social consequences of which are not even understood, does pose the danger of setting us on the path to premature and ill-considered action against enhancements. Recall that Mehlman *does* call for a ban to be implemented on germline enhancements because he believes they are simply too dangerous; they will create a genetic aristocracy. However, the seizure of power by a genetic upper class is highly improbable. There are many obstacles to such a radical transformation of society. For example, unless one assumes the unenhanced will cede power without a struggle, the genetic aristocracy will have to seize power surreptitiously to establish their domination. That contingency requires the increasing disparity between enhanced and unenhanced to pass unnoticed, which is unlikely. Mehlman is calling for a preemptive strike against germline enhancements based on nothing more than speculation.

We have seen what preemptive strikes based on guesswork have brought us in terms of foreign policy disasters, but at least with respect to Bush's invasion of Iraq, we had some understanding of what harm chemical, biological, and nuclear weapons can cause—it just turned out that Saddam Hussein did not have any. Trying to devise a plan for a society transformed by enhancements, especially to the extent these plans call for a ban on some enhancements, when we have no understanding of what the societal effects of these enhancements might be is highly inadvisable. Consider where we would be if circa 1935, when the first experiments in controlling ovulation in animals were carried out, we had decided that oral contraceptives for humans should be banned because their availability would cause too much social unrest, or, alternatively, women who had the disposable income to purchase oral contraceptives should be required to sign chastity pledges and donate time to helping others raise their children, because these women would enjoy special privileges and in exchange for these privileges they had an obligation to "abide by rules designed to promote social goals and to refrain from behaving in socially undesirable ways."[52] This hypothetical example is not as farfetched as it seems. Until 1965, some states could and did ban use of contraceptives.[53] Trying to devise a plan at present to ensure a just society in an era of enhancements is equivalent to having provided the citizens of ancient Athens the authority to establish rules governing the development of any future technological innovations.

The problem with devising plans for the future is that we don't know what it will be like, especially if—and this is the assumption of those who are concerned about enhancements—the future will be radically different from our present situation.

Mehlman, as indicated, has suggested that I am indifferent about the effects of enhancements on our society. Not so. But for the reasons indicated, and some additional ones set forth below, I oppose regulating or banning enhancements now.

First, the assumption that only the wealthy will have access to enhancements, although not entirely unreasonable, is far from certain. If we look at history, we see that many technological innovations, from telephones to cars to televisions to digital cameras, were often affordable initially only by the wealthy. Within a matter of a few years, they were reduced sufficiently in price that millions of people could purchase them. Economics and market competition play a role here. Selling the IntelliPill at $100,000 per annual dose will attract only so many customers. Reducing this enhancement to $1,000 per dose will generate much more income for the developer. By contrast, the best way to guarantee that only the wealthy will have access to enhancements is to ban them. If you haven't noticed, people with lots of money often find ways to circumvent the law.

Second, assuming only the wealthy will have access to enhancements, is there something horribly wrong about that prospect? Does justice require that wealthy individuals spend their excess money on yachts, mansions, luxurious cars, and three-month vacations, instead of enhancements to improve verbal comprehension, for example? Or does justice require wealthy persons to spend their money on "normal" private education instead of an enhancement, even if the education will have little effect without the enhancement? Why? Because we think that enhanced verbal skills *might* threaten us in ways that the current ability of the wealthy to influence public policy does not?

This leads to my next point. Mehlman and others are concerned about the threat enhancements pose to our "liberal democracy." Our current political and social arrangements are probably preferable to those of, for example, France in 1750. But our "liberal democracy" is one in which the wealthy have enormous power. Here's a contest for you: you and Bill Gates both call your congressional representative at the same time. See whose call gets answered first.*

Gaps in wealth and income are both a source and an object of concern for those troubled by enhancements. So why not deal with these dispari-

*If she buys this book, Melinda Gates is disqualified from participating in this contest.

ties directly through progressive income taxation, capital gains taxes, and inheritance taxes, rather than indirectly through restriction of enhancements? Redistribution of wealth is morally less problematic than stringent restrictions or bans on enhancements. Wealth redistribution restricts a person's liberty much less than prohibiting her from pursuing improved intellectual or physical skills. For virtually everyone affected by it, a modest redistribution of income or wealth does not preclude pursuit of worthwhile activities, such as using one's enhanced intelligence to study molecular biology and, perhaps, devise an inexpensive way to produce enhancements for all. Moreover, redistribution of financial resources helps others to achieve their goals by giving them the means to do so. By contrast, prohibiting the use of enhancements may deprive the financially fortunate of the most effective, if not the only, means they may have to pursue some worthwhile activities while doing *nothing* to assist those who lack the resources to do so. Where is the justice in that? In addition, for all the reasons already set forth, an equitable redistribution of wealth would be much easier to formulate and manage than any attempt to control, manage, and distribute enhancements.

The irony is that during the last couple of decades, during which time there has been much agonizing over the possible effects of enhancements, the American tax system has become much less progressive and the gaps in income and wealth between the very rich and everyone else have increased. For example, the wealthy and their bought politicians have even been able to persuade many that the estate tax, which is a useful (but far from completely effective) means of preventing the accumulation and concentration of intergenerational wealth, is a "death" tax that should be abolished. While our attention has been focused on the scaremongers who warn us of the horrors of a "posthuman" future, our pockets have been picked.

No, I don't think there is any conspiracy between the wealthy who dominate our politics and those who decry the supposedly dehumanizing or unnatural consequences of the biotech revolution. On the other hand, one way to help ensure that the social and political status quo is preserved is to prevent the masses from improving themselves. Even an enhancement as limited as one that increases a person's analytical ability slightly would pose a threat to those whose power rests on the effectiveness of political spin and the gullibility of the majority of the populace. If a liberal democracy functions better when its citizens are able to make informed, rational decisions, then banning enhancements seems more likely to impair liberal democracy than promote it.

I do not deny that we will be required to confront many novel situations if and when significant pharmacological and genetic enhancements become available. However, we certainly will be in a better position to evaluate the consequences of enhancements *after* they take place, and there is little reason to suspect we will be unable to take appropriate action to manage these consequences. So far, we have been able to handle issues presented by new technologies as they arise. On the other hand, there is reason to be concerned that impulsive and premature rejection of new technologies will deprive us of many benefits. One should not need any special cognitive enhancement to realize that.

Of course, without adequate research, we may never have the opportunity to develop enhancements of any sort. Scientific advances depend on research. There are many stringent regulations in place that govern and guide research dealing with human subjects. No one questions the importance of such regulations. However, in the last few years there has been enormous controversy over research involving embryos. Some have argued that such research is unethical, and, at present, there is a ban on federal funding of research that "harms" embryos. Is there a justification for such a prohibition? Should that prohibition be modified to allow for embryonic stem cell research? We will address these issues in the next chapter.

NOTES

1. See, for example, Francis Fukuyama, *Our Posthuman Future: Consequences of the Biotechnology Revolution* (New York: Farrar, Straus and Giroux, 2002), pp. 208–209.

2. Allen Buchanan et al., *From Chance to Choice* (Cambridge: Cambridge University Press, 2000), p. 321.

3. Maxwell J. Mehlman, *Wondergenes* (Bloomington: Indiana University Press, 2003), pp. 119–20.

4. Ibid., pp. 155–91.

5. Michael J. Sandel, *The Case against Perfection: Ethics in the Age of Genetic Engineering* (Cambridge, MA: Harvard University Press, 2007), p. 16.

6. Ibid., p. 87.

7. Leon Kass, "Ageless Bodies, Happy Souls: Biotechnology and the Pursuit of Perfection," *New Atlantis*, Spring 2003, pp. 9–28.

8. George J. Annas, "The Man on the Moon, Immortality, and Other Millennial Myths: The Prospects and Perils of Human Genetic Engineering," *Emory Law Journal* 49 (2000): 779–80.

9. Joe Z. Tsien, "Building a Brainier Mouse," *Scientific American*, April 2000, pp. 62–68.

10. Human Genome Project Information, "What Is Gene Therapy?" US Department of Energy Office of Science, available at http://www.ornl.gov/sci/techresources/Human_Genome/medicine/genetherapy.shtml (accessed October 7, 2007). See also Rick Weiss, "Fungus Infected Woman Who Died after Gene Therapy," *Washington Post*, August 17, 2007, p. A10.

11. See, for example, Philip Kitcher, "Creating Perfect People," in *A Companion to Genethics*, ed. Justine Burley and John Harris (Oxford: Blackwell, 2002), pp. 229–42.

12. For a discussion of the cognitive effects of modafinil, see Danielle C. Turner et al., "Cognitive Enhancing Effects of Modafinil in Healthy Volunteers," *Psychopharmacology* 165 (2003): 260–69. For a discussion of the use of modafinil for nontherapeutic purposes, see David Plotz, "Wake Up, Little Susie: Can We Sleep Less?" *Slate*, March 7, 2003, available at http://www.slate.com/id/2079113 (accessed October 7, 2007).

13. Council of Europe, *Convention for the Protection of Human Rights and Dignity of the Human Being with Regard to the Application of Biology and Medicine: Con-*

vention on Human Rights and Biomedicine, ETS no. 164 (Oviedo: 1997), chap. 4, art. 13. This treaty is available at http://conventions.coe.int.Treaty.en/Treaties/Word/164.doc (accessed October 18, 2007).

14. Southern Baptist Convention, On Human Species-Altering Technologies, Resolution no. 7 (Greensboro, NC: 2006), available at http://www.sbcannualmeeting.net/sbc06/resolutions/sbcresolution-06.asp?ID=7 (accessed October 18, 2007).

15. See, for example, Norman Daniels, *Just Health Care* (Cambridge: Cambridge University Press, 1985).

16. H. Tristram Engelhardt Jr., "The Disease of Masturbation: Values and the Concept of Disease," *Bulletin of the History of Medicine* 48 (Summer 1974): 234–48.

17. For a discussion of Lesch-Nyhan syndrome, see Richard Preston, "An Error in the Code," *New Yorker*, August 13, 2007, pp. 30–36.

18. Claire Tomalin, *Thomas Hardy* (New York: Penguin Press, 2007), p. 42.

19. Sandel, *The Case against Perfection*, pp. 26–27.

20. Ibid., p. 27.

21. Ibid., pp. 93–95.

22. Ibid., p. 39.

23. Ibid., p. 31.

24. See, for example, Ted D. Adams et al., "Long-Term Mortality after Gastric Bypass Surgery," *New England Journal of Medicine* 357 (2007): 753–61.

25. Sandel, *The Case against Perfection*, p. 8.

26. Kass, "Ageless Bodies, Happy Souls," p. 22.

27. Ibid., p. 23.

28. I should note that Erik Parens, a scholar who has written widely on enhancements, takes direct issue with my example. He argues that improving a child's education by reducing classroom size is categorically different than improving a child's ability to learn by providing the child with a drug such as Ritalin. To quote Parens, "[O]ne experience entails reduced 'noise' in the child's brain, the other reduced noise in her classroom." See his "Is Better Always Good? The Enhancement Project" in *Enhancing Human Traits: Ethical and Social Implications*, ed. Erik Parens (Washington, DC: Georgetown University Press, 1998), p. 13. Parens overlooks the fact that the reduced noise in the classroom is effective in improving the child's education only because of the impact it has on the child's brain. Both drug and classroom size affect the child's brain.

29. Kass, "Ageless Bodies, Happy Souls," p. 24.

30. Fukuyama, *Our Posthuman Future*, p. 46.

31. Sandel, *The Case against Perfection*, pp. 25–44. See also Erik Parens, "The Goodness of Fragility: On the Prospect of Genetic Technologies Aimed at the Enhancement of Human Capacities," *Kennedy Institute of Ethics Journal* 5 (1995): 141–53.

32. This is an allusion, of course, to the notorious case in which Bill Belichick, the coach of the New England Patriots, was fined $500,000 for having the signals of an opposing football team, the New York Jets, videotaped. Given the relative strengths of the Patriots and the Jets in the 2007 season, I'm not sure this was money well spent. See Judy Battista, "Sideline Spying: N.F.L. Punishes Patriots' Taping," *New York Times*, September 14, 2007, p. A1.

33. Kass, "Ageless Bodies, Happy Souls," pp. 16–17.

34. Parens, "Is Better Always Good?" p. 15.

35. Bill McKibben, *Enough: Staying Human in an Engineered Age* (New York: Times Books, 2003), p. 191.

36. Sandel, *The Case against Perfection*, pp. 45–49.

37. See Seth Mydans, "After 5 Months of Drama, Brothers' Trials Near End," *New York Times*, December 12, 1993.

38. Sandel, *The Case against Perfection*, p. 87.

39. Ibid.

40. George J. Annas, Lori B. Andrews, and Rosario M. Isasi, "Protecting the Endangered Human: Toward an International Treaty Prohibiting Cloning and Inheritable Alterations," *American Journal of Law and Medicine* 28 (2002): 153.

41. Kass, "Ageless Bodies, Happy Souls," p. 17.

42. Buchanan et al., *From Chance to Choice*, pp. 95–96.

43. For a more in-depth account of the problems with devising principles of justice for a society transformed by enhancements, see my "Enhancements and Justice: Problems in Determining the Requirements of Justice in a Genetically Transformed Society," *Kennedy Institute of Ethics Journal* 15 (2005): 3–38.

44. Maxwell J. Mehlman, "The Law of Above Averages: Leveling the New Genetic Enhancement Playing Field," *Iowa Law Review* 85 (2000): 577.

45. Ibid., p. 570.

46. Ibid., p. 571.

47. Ibid.

48. Mehlman, *Wondergenes*, p. 179.

49. I have been a lawyer for nearly three decades.

50. Mehlman, "The Law of Above Averages," pp. 576–77.

51. Maxwell J. Mehlman, "Plan Now to Act Later," *Kennedy Institute of Ethics Journal* 15 (2005): 81.

52. Cf. Mehlman, "The Law of Above Averages," p. 571.

53. In *Griswold v. Connecticut*, 381 U.S. 479 (1965), the Supreme Court invalidated Connecticut's criminal ban on contraception.

7.

Saving Embryos for the Trash—Our Illogical Policies on Embryonic Stem Cell Research

Perhaps no bioethical controversy in recent times better illustrates the power of misguided thinking to impede progressive policies than the controversy over the funding of embryonic stem cell research by the federal government.* Those who oppose such funding offer arguments that rest on bad science and are internally inconsistent. Nonetheless, these arguments have attracted sufficient support to allow the Bush administration to block regulations and legislation that would permit federal funding of embryonic stem cell research.

Research on stem cells, embryonic and adult, as well as stem cells derived from amniotic fluid, reprogrammed somatic cells, or other sources, is evolving rapidly. It seems as though every few months a new finding is announced. Occasionally, a new technique is heralded as a means of avoiding the acrimonious debate over embryonic stem cell research.[1] Accordingly, it is conceivable that by the time this book is published, there will be some development not addressed in these pages. But the fundamental policy issues will remain, for reasons that will become

* Throughout this chapter, I will be using the term "embryo" as it is traditionally used, that is, to designate an organism in the first two months of its development following formation of a zygote (the one-celled fertilized egg that has formed a nucleus). After two months the organism is usually referred to as a fetus. In some contexts, I may refer to different stages of the embryo (for example, the blastocyst stage) as needed.

clearer later in this chapter. For now, let me just mention that almost all scientists agree that so-called alternatives to embryonic stem cell research will at most supplement this research, not supplant it.

Let's review some of the history of the controversy first. I will then discuss the basic science behind embryonic stem cell research. Then, in accordance with my established procedure, I will set forth some basic moral and factual premises that will guide us in our policy discussion.

CHRONOLOGY OF THE EMBRYONIC STEM CELL FUNDING DISPUTE

Stem cells are unspecialized cells that have the capacity to produce more stem cells and also all the different types of cells in the body, for example, liver cells or neurons. (For this reason, they are called "pluripotent.") As discussed in more detail below, one reason research on stem cells is considered important is that it has the potential to allow us to develop effective therapies for all sorts of illnesses and injuries. If stem cells can be reliably directed to differentiate into specific cell types, there is the possibility of developing replacement tissues for millions of Americans who suffer from debilitating diseases and disabilities, including Parkinson's and Alzheimer's diseases, diabetes, heart disease, liver disease, and spinal cord injury, to name just a few.

In 1998, researchers at the University of Wisconsin succeeded in isolating human embryonic stem cells and cultivating stem cell lines based on these cells.[2] In other words, the scientists succeeded in getting the cells to continue to live and divide, generating more stem cells. Furthermore, when injected into mice, the stem cells differentiated into various types of tissue.

The importance of this development quickly became apparent. As one might expect, the federal government also had to decide whether to support research on embryonic stem cells. Since the late nineteenth century, the federal government has recognized as part of its mission the funding of medical research that could assist the people of the United

States. The National Institutes of Health (NIH), the primary government body overseeing medical research in this country, spends between $25 billion and $30 billion in research each year. One problem in funding research on embryonic stem cells is that it arguably would violate the so-called Dickey Amendment, a provision added to a budget bill in 1996, which prohibits federal funding of research that results in the destruction of embryos or subjects embryos to a risk of injury.[3] This provision had been adopted prior to the cultivation of human embryonic stem cells, and it was not entirely clear how it affected funding of such research. After some deliberation, the federal government decided that it could fund research on lines established by private research entities (universities or companies), as it is the creation of the line that arguably results in the destruction of the embryo, not cultivation of the line. In August 2000, (NIH) issued guidelines that would have allowed federal funding of embryonic stem cell research.[4] Then President Bill Clinton supported the guidelines.

However, soon after taking office, President George W. Bush placed an indefinite hold on federal funding while his administration reviewed the policy. After several months of deliberation and his own personal consultation with various hand-picked scholars, on August 9, 2001, in his first televised address to the nation, Bush announced that federal funding would be provided only for those few dozen embryonic stem cell lines already in existence. He claimed they were sufficient to allow meaningful research to take place.[5] He admitted that there was some difference of opinion among the scholars he consulted, but in striking what he recognized was a compromise, he explained that "human life is a sacred gift from our Creator" and he worried about a "culture that devalues life."[6] Implicit in Bush's remarks was the claim that an embryo was equivalent in moral status to an adult human being. This claim later became explicit.

As things have turned out, the stem cell lines that Bush had stated would be sufficient for research are inadequate for a meaningful research program. To begin, instead of the sixty stem cell lines mentioned in Bush's address, only about twenty lines have been available for distribution to researchers by NIH. The other lines have chromosomal abnormalities or other problems. In addition, as cell lines age, they become more susceptible to abnormalities. Thus, over time all these cell lines will become unusable. Finally, the approved cell lines were created through a process that utilized mouse cells (typically, as "feeder" cells that supplied nutrients to human embryonic stem cells). Studies have confirmed that the stem cells from these lines contain mouse proteins on their surface. Effectively, this means cells derived from these lines will never be available for therapeutic purposes because they would trigger an immune reaction. Newer techniques can create embryonic stem cell lines without the use of animal cells.

Objections to the Bush administration policy were made very soon after it was announced, but given the substantial support the president enjoyed in Congress, no serious effort was made to overturn his ban until 2005. Then legislation was introduced to allow funding of research carried out on a limited class of embryonic stem cells, namely, those derived from spare embryos generated through in vitro fertilization (IVF) procedures. In IVF procedures, for reasons of time and efficiency, typically far more embryos are created than are actually needed for implantation. Most of these spare embryos are eventually discarded. Estimates are that there are about four hundred thousand spare embryos currently available in the United States.[7] Both the House and the Senate passed the proposed legislation by July 2006.

President Bush vetoed the legislation, characterizing embryonic stem cell research as "the taking of innocent human life" and asserting that each embryo "is a unique human life with inherent dignity and matchless value."[8] The president's press secretary at the time, Tony Snow, stated that the destruction of embryos during research was "murder," although he later explained he had "overstepped" his brief and that Bush himself would not use the term "murder," even though the research results in the "destruction of human life."[9] Apparently, although it supposedly involves the "taking of innocent human life," using embryonic stem cells in research is not quite murder. Perhaps it is voluntary manslaughter?

Similar legislation to allow research on spare embryos from IVF procedures was passed by Congress again in 2007, and again, the president vetoed the legislation.[10] It is evident that there will be no federal funding of stem cell research at least as long as Bush remains in the White House.

Bush's opposition to embryonic stem cell research is based on the premise that embryonic stem cell research destroys the embryo and the destruction of the embryo is equivalent to the intentional killing of a human being (although perhaps not murder). If human embryos are entitled to the full protection of our moral norms and the use of such embryos in research constitutes killing, then opposition to such research is understandable. However, we should not simply assume that our moral norms and principles apply to embryos. At a minimum, we need a well-reasoned and compelling argument, firmly grounded in scientific evidence, to support an extension of our moral norms and principles to encompass embryos. As discussed in more detail below, to interpret norms that prohibit unjustified killing so they also prohibit the use of embryos in research leads to many difficulties, paradoxes, inconsistencies, and morally indefensible conclusions. Some of the difficulties in articulating a coherent position against embryonic stem cell research are illustrated by Tony Snow's fumbling for the correct term to describe Bush's understanding of what researchers do when they use the embryo. Unless excused by self-defense or some comparable justification, intentional killing of a human being constitutes some form of criminal homicide, but even opponents of embryonic stem cell research shrink from using such terms. Why? Well, as I will demonstrate below, one reason they hesitate to describe the use of embryos in research as murder is that in other contexts hardly anyone regards embryos as possessing the status of a human person. Moreover, arguments to assign embryos the status of human persons ultimately founder on both biology and logic.

Let us now discuss the science behind embryonic stem cell research.

SCIENTIFIC BACKGROUND FOR STEM CELL RESEARCH

A basic understanding of the science of stem cells and embryonic development is necessary before discussing the ethical implications of stem cell research.

Stem cells are present during all stages of an organism, including the embryonic, fetal, and adult stages, but the characteristics of stem cells at various stages differ.[11]

In particular, embryonic stem cells have properties that are different from fetal or adult stem cells. In the early embryo (to around the five-day stage), each cell is totipotent, that is, under the appropriate conditions each cell could develop into a complete, individual organism. After five days, the embryo becomes a blastocyst, an entity of about one hundred to two hundred cells that consists of an outer sphere that can develop into membranes, such as the placenta, and an inner cell mass that can develop into a fetus. The cells of the inner mass are pluripotent, but they are no longer totipotent. As indicated previously, "pluripotent" means the cells can develop into all the different cell types in the body. Embryonic stem cells from the inner mass of the blastocyst stage are currently the preferred source of stem cells for research. One reason they are used in research is that because of their pluripotent capabilities, and their ability to proliferate under laboratory conditions, they are considered to be more advantageous for work on most research projects than fetal stem cells and adult stem cells.[12]

One problem with adult stem cells is that they do not appear to have the same potential to proliferate under research conditions as embryonic stem cells. Embryonic stem cells can proliferate for a year or more in the laboratory without differentiating, but to date scientists have been unsuccessful in obtaining similar results with adult stem cells. This distinction is important because to develop therapies large numbers of stem cells would be needed. Moreover—although the research on this issue is ongoing—adult stem cells appear, at best, to be multipotent or multisomatic rather than pluripotent. It is unclear whether true transdifferentiation can occur with adult stem cells, that is, it is unclear whether adult stem cells have the ability to develop into many different types of tissue as opposed to developing into different types of cells of similar tissue.[13] Recent research has also revealed that adult stem cells lack a key protein that maintains the pluripotency of embryonic stem cells, which at least suggests they are not pluripotent.[14] Nonetheless, there is some evidence that hematopoietic stem cells (a type of stem cell found in bone marrow) not only can develop into all types of blood cells, but can also develop into neurons and cardiac muscle cells.[15]

One possible major advantage of adult stem cells is that if replacement tissues could be developed from a person's own adult stem cells, there would not be a concern about an immune reaction, whereas there would be such a concern with respect to replacement tissues developed from embryonic stem cells.

Most fetal cells have limitations similar to adult stem cells, that is, they appear to be limited in their ability to be transformed into different cell types. Moreover, there are ethical objections to using in research fetal stem cells extracted from a fetus, so it is doubtful whether, in terms of the public policy debate, there is any advantage to using such cells in research instead of embryonic stem cells. However, fetal cells that are shed in the amniotic fluid that surrounds the developing fetus have been shown, under some circumstances, to have the capacity to be pluripotent.[16] They also appear to have the ability to proliferate as well as embryonic stem cells. This could be an important development. If research confirms the potential of fetal stem cells derived from amniotic fluid to be truly pluripotent, then we would have a promising set of stem cells that could be used for research purposes. Moreover, it appears that the initial reaction of many who are opposed to embryonic stem cell research is that they would not oppose research with fetal cells derived from amniotic fluid because they could be collected without harming the fetus.[17]

There are some variants on the three types of stem cells discussed above—embryonic, fetal, adult—but to understand these variants, one has to understand a bit more about the techniques for producing stem cells, so I turn to that discussion.

At the moment, IVF is the most reliable way to produce an embryo that could yield stem cells. IVF, of course, is designed to assist a woman who, for whatever reason, is unable to conceive and bear a child otherwise. Many embryos produced as a result of IVF end up being spare embryos that will not be implanted in a uterus. The decision not to use them can be based on a number of different reasons, including the fact that the woman has already achieved a successful implantation with another embryo or the presence of detected abnormalities in the embryo. Obviously, embryos could also be created through IVF for the express purpose of using them for research. However, the legislation passed by Congress that would have allowed federal funding expressly forbids the use of funds for research on embryos created solely for research purposes. Funding was limited to research on embryos created for the purpose of fertility treatment.

With respect to extracting a cell that can be used to create a stem cell line, the standard procedure is to separate the inner cell mass of the blastocyst from the outer sphere of cells and then have this inner cell mass cultured on a plate of feeder cells that will maintain the stem cells through a supply of nutrients. After the cells of the inner cell mass begin to proliferate, they are removed and plated into fresh culture dishes and, eventually, if the process is successful, an embryonic stem cell line will be established. This is the process that opponents of embryonic stem cell research characterize as destroying or killing the embryo. Technically, this is incorrect. The embryo (because it has had its outer cell mass removed) will not develop into a fetus—unless it is given a placenta and some other assistance (more on this below). On the other hand, its cells remain alive. Indeed, they may remain alive far longer than they would have if the embryo had continued its development. The vast majority of embryos created through standard sexual reproduction never become children, which is a fact important for other reasons, as I will discuss below.

In 2006, researchers announced that another procedure had been used successfully. In this procedure, a stem cell line is developed from the cells of an embryo at a very early (eight-cell) stage, when the individual cells are known as "blastomeres."[18] When this news was announced, some suggested that this procedure could obviate the controversy over embryonic stem cell research because when a blastomere is removed from an embryo for pre-implantation genetic diagnosis (a procedure commonly used in connection with IVF), the remaining seven cells of the embryo typically have *not* been prevented from maturing and differentiating into a fetus and then an infant. In other words, the embryo is not "killed."

However, it is too early to determine what effect this new procedure will have on the debate over embryonic stem cell research. The initial attempt to use this new procedure to generate a stem cell line actually prevented the embryo from developing further. At the time of this writing, researchers recently announced that they have been successful in developing a stem cell line from blastomeres without halting the maturation of the remaining cells. Therefore, at least arguably, stem cell lines can be developed without "killing" the embryo.[19] Nonetheless, those who believe the embryo has the status of a human person have already stated that they would object to such a procedure because of its potential for harming the embryo, whether or not it actually precludes maturation of the embryo.[20] Recall also that the Dickey Amendment prohibits research that subjects an embryo to a "risk of injury." Therefore, this new procedure does not appear to be a way to avoid the policy disputes over embryonic stem cell research.

In addition to creation of embryos through IVF, there is the possibility of creating embryos (or would they be true embryos?) through somatic cell nuclear transfer (SCNT). Do you remember Dolly, the cloned sheep? This is the procedure that was used to create her and the other mammals that have been successfully cloned. SCNT is accomplished by culturing the nucleus of a somatic cell and then transferring this nucleus into an enucleated ovum (an egg cell that has had its nucleus removed). This new cell is then stimulated to divide and, when the procedure works, the cell will develop into a blastocyst with the genotype of the somatic cell donor (but with the mitochondrial DNA of the ovum). Were scientists to develop a human embryo with this procedure, they could then extract the stem cells using one of the two procedures outlined above.

In January 2008, Samuel Wood of Stemagen, a privately held company, announced that his company had succeeded in creating cloned human embryos.[21] This announcement was greeted with caution because of prior claims by other researchers that later proved spurious. You may recall that a few years ago a Korean scientist claimed to have developed a cloned human embryo, but this claim was later discovered to be fraudulent. Moreover, it is unclear whether Stemagen's cloned embryos were "healthy," that is, that they could develop into fetuses and possibly children, under the appropriate conditions. Stemagen destroyed the embryos it created after a few days. Some have argued that no human embryo created through SCNT could be successfully brought to term.[22] Essentially, without getting into all the technical details, the argument is that in the process of removing the nuclear material from human eggs, proteins essential for embryonic development may also be removed. If it turns out that humans cannot be reproduced through cloning, think of all the science fiction movies and books that will seem outdated! But it's still too early to draw any firm conclusions. The success rate for reproductive cloning of any mammal is not very high.

But precisely because an embryo created through SCNT may not be able to be carried to term, some have maintained that this procedure can ethically be used to create embryos for research purposes. Merely because they cannot be carried to term does not entail that we would not be able to cultivate a stem cell line out of such embryos. Some have even suggested that an egg that has received its nucleus from a somatic cell should be called a "clonote" as distinguished from zygote, which is the term used to describe an egg successfully fertilized by sperm.[23]

This is an appropriate place to discuss two other possible methods of creation of stem cells that arguably would not involve embryos, although one of them at least involves an "embryo-like" entity. These methods have been proposed as ways of avoiding the supposedly difficult moral questions presented by using embryonic stem cells. Both of these methods are discussed in a white paper prepared by the President's Council on Bioethics and have been touted by some opponents of embryonic stem cell research.[24] The first method is referred to as "altered nuclear transfer."

It is a variant on SCNT. In altered nuclear transfer, *before* the nucleus of the somatic cell is transferred into the ovum, it is damaged in some way by removing or suppressing a gene essential for standard embryogenesis. The thought behind this procedure is that if genes necessary for proper embryonic development will be altered or removed, the resulting "artifact" will not be a true embryo, but, nonetheless, will be capable of generating stem cells. However, not only is this proposed technique mere speculation at this stage, some see it as posing even more serious moral problems than embryonic stem cell research. If the embryo is the equivalent of a human, as some claim, then this procedure could be characterized as intentionally creating severely damaged humans. Again, bear in mind that most embryos created through standard reproduction (the old-fashioned sperm-meets-egg method) never develop completely, and most scientists believe this is because they are defective in some way. However, they are still considered embryos.

Somatic cell "dedifferentiation" would take a regular somatic cell, such as one of our skin cells, and reprogram it to return it to a state where it would become a stem cell. In other words, it would effectively reverse the process whereby stem cells become different types of cells; hence the term "dedifferentiation." In November 2007, separate teams of scientists in Japan and Wisconsin announced that they had managed to induce human skin cells to reprogram themselves into cells having the characteristics of embryonic stem cells.[25] (The Wisconsin team was led by James Thomson, who had been the first to develop stem cell lines from human embryos.) Through prior work with mice, the scientists had identified a set of four genes that turn other genes in a cell "on" or "off." By inserting designed viruses carrying these genes into the human skin cells, the scientists were able to control the regulatory genes in a few instances (one in five thousand cells) and induce the somatic cell to acquire the pluripotent characteristics of an embryonic stem cell. Hence these cells are called induced pluripotent stem cells, or iPS cells. The announcement that scientists had developed iPS cells was appropriately hailed with enthusiasm. However, there are a few sizable kinks remaining to be worked out. For example, when mice were injected with iPS cells, they developed cancer.

Still, there is no question that iPS cells will be valuable in research. Whether iPS cells can obviate the need for "regular" embryonic stem cells is another question, to be discussed below.

Finally, let me discuss embryos and embryonic development a bit. The picture some people have of an embryo is that it is a human being in miniature. If left on its own, it will naturally grow into someone who might buy the third edition of this book. The biology is a bit more complicated than this.

First, an embryo prior to gastrulation, that is, the point in time when the so-called "primitive streak," the precursor to the spinal cord, appears, may develop into more than one individual. This is the phenomenon known as "twinning." Another process works the other way. Women sometimes ovulate two eggs, and these eggs can be fertilized by different sperm. Two separately fertilized eggs typically develop as fraternal twins. However, at a very early stage of embryo development, before any differentiation has taken place, they can also fuse into one individual, known as a "chimera." This one individual will carry the DNA of both fertilized eggs. These examples are important because they at least suggest that the embryo does not become a true individual with a determinate identity until gastrulation, which happens around the fourteenth day after fertilization.[26]

One final fact, which will underscore how biology tends to undermine the preconceived notions some have. You will recall that currently the standard process for obtaining embryonic stem cells involves separating the inner cell mass of the blastocyst from the outer sphere of the blastocyst. This prevents further embryonic development—but this result is not necessarily inevitable. Scientists have determined how embryonic stem cells can be coaxed into developing placental cells.[27] So it is possible that one could isolate some stem cells, work with them, cause some of them to form a placenta, implant the newly formed (restored?) embryo into a uterus, and have a baby appear in nine months. Thus, at least in theory, embryonic stem cells could become adult human beings under the right conditions. I'll discuss the moral implications of this below.

MORAL AND FACTUAL PREMISES FOR RESOLVING THE DISPUTE OVER FEDERAL FUNDING

I will take it as given that the federal government should support medical research that has a reasonable prospect of providing important benefits to many people. Hardcore libertarians may disagree with this premise, but this premise is accepted, I believe, by the vast majority of people in the United States. Obviously, support for research has to be balanced against other needs, but in the dispute over embryonic stem cell research, the debate has not been over the amount of money that should go to such research, but whether any money should go to such research.

Embryonic stem cell research does have the potential to provide us with revolutionary therapies. Although there is no certainty that such therapies could be developed—and, realistically, they are at least a couple of decades away—the research to date appears promising. For example, dopamine-producing neurons generated from mouse embryonic stem cells have proved functional in animals, thus indicating there is a realistic possibility that similar results could be reproduced in humans, with beneficial consequences to those suffering from Parkinson's.[28] Even more dramatic results were obtained via an experiment conducted by researchers at Johns Hopkins University. These researchers were able to use neurons derived from embryonic stem cells to restore motor function in paralyzed rats.[29]

Research on embryonic stem cells is not limited to trying to grow them into replacement tissues. Embryonic stem cells are also a critical tool in learning about early human development, including the causes of birth defects. Understanding the chemical signals that cells transmit is essential for exploiting the potential of any stem cell, whether it is embryonic, fetal, or adult. We need embryonic stem cells to advance our knowledge in this area.

Potential benefits from scientific experiments cannot, of course, by themselves justify scientific research. We do not force adult humans to ingest drugs that are being developed merely because it would be helpful in advancing our scientific knowledge.

This brings us to the key objection to embryonic stem cell research. That is the claim that such research is impermissible because it harms, destroys, or "kills" an entity that is the equivalent of a human person. Arguments about the moral status of an entity are notoriously difficult to resolve. Consider the disputes, for example, about what rights a fetus has or the rights of animals. Indeed, I do not believe disputes about moral status can be resolved merely through consideration of the norms of common morality, that is, the moral norms nearly all morally serious people accept and that I have relied on throughout this book to help us in resolving policy disputes. It is true that throughout history, all human cultures have accepted it is wrong to kill another member of the moral community, absent some justification such as self-defense. But deciding who is a member of the moral community is a separate question and lies outside the scope of the common morality.[30] In fact, whereas today most of us believe that all humans, or at least all humans who have the capacity to be considered moral agents, are members of the moral community, this has not always been the prevailing view. Quite to the contrary, in most cultures prior to modern times, "barbarians," "infidels," and other outsiders were not members of one's moral community and were not entitled to moral respect. They could be enslaved or, in some cases, killed with impunity. If you have any doubt about this, consider the reaction of Moses, the person who gave us the Ten Commandments and one of the acknowledged teachers of the common morality, after he learned that the Israelites had taken some women and children as prisoners after attacking another tribe.* Moses was quite angry with this show of mercy to the ungodly and directed as follows:

> Now therefore, kill every male among the little ones, and kill every woman who has known man by lying with him. But all the young girls who have not known man by lying with him, keep alive for yourselves.[31]

---

* Moses may well be a legendary figure, as opposed to an actual person. That question is irrelevant to the point I am making.

We may agree with Moses and those portions of the Ten Commandments that instruct us that it is wrong to kill, steal, or lie. In other words, just as he did, we accept the norms of the common morality. Presumably, however, we would disagree with him that these norms do not apply to other humans, and we can kill them or convert them into sex slaves, merely because they have different religious beliefs. The norms of the common morality tell us what to do or refrain from doing; they do not tell us to whom these norms apply.

But does that mean we have no way of resolving the dispute over the status of embryos? No. Recall from chapter 2 that I stated we can test moral claims by determining whether they are supported by scientific evidence, by considering their implications to see if they are consistent with other moral beliefs, and by evaluating them for coherence. Claims that the embryo is the equivalent of an adult human being fail these tests. It is also important that we do not lose sight of the fact that morality is a practical enterprise. We need to understand the rationale of our moral norms if we are to apply them successfully. In deciding whether the embryo is entitled to moral consideration equal to that extended to human persons, we need to ask ourselves whether such treatment serves the objectives of morality. Unfortunately, this is a question that few have paused to consider in the debate over embryonic stem cell research.

THE POSITION THAT EMBRYOS ARE ENTITLED TO THE SAME RIGHTS AS HUMAN PERSONS

There are some who believe that embryos deserve the full range of rights provided to human persons and that removing from an embryo the possibility of developing the capacities and properties characteristic of human persons is morally equivalent to killing an adult human. Those who hold this view maintain we should not "harm" embryos by utilizing them in stem cell research, just as we do not kill adult humans for research purposes.

An essential premise of this position is that even though the embryo does not currently possess the capacities and properties of human persons, it possesses the potential to develop these capacities and properties, and this potential is sufficient to provide it with the moral status of a human person. On this view, an embryo is merely a human person at an early stage of development. Another essential premise of this position—but one that is not always acknowledged—is that the embryo is already an individual even at its earliest stages of development. To claim that someone is harmed, there must be "someone" there. As the President's Council on Bioethics recognizes, "individuality is essential to human personhood and capacity for moral status."[32] We do not grant moral rights to mere groupings of cells, even if they are genetically unique.

The argument that the embryo is entitled to the same rights as human persons has been most clearly articulated and ably defended by Professors Robert George and Alfonso Gómez-Lobo, two members of the President's Council on Bioethics. In a statement that appears in the council's 2002 report entitled *Human Cloning and Human Dignity*, George and Gómez-Lobo assert that:

> A human embryo is a whole living member of the species *Homo sapiens* in the earliest stage of his or her natural development. . . . The embryonic, fetal, infant . . . stages are stages in the development of a determinate and enduring entity—a human being—who comes into existence as a single cell organism and develops, if all goes well, into adulthood many years later.
>
> Human embryos possess the epigenetic primordia for self-directed growth into adulthood, with their determinateness and identity fully intact. The adult human being that is now you or me is the same human being who, at an earlier stage of his or her life, was an adolescent, and before that a child, an infant, a fetus and an embryo . . . .

Of course, human beings in the embryonic, fetal, and early infant stages lack immediately exercisable capacities for mental functions characteristically carried out . . . by most . . . human beings at later stages of maturity. Still, they possess in radical (=root) form these very capacities. . . . As humans, they are members of a natural kind—the human species—whose embryonic, fetal, and infant members, if not prevented by some extrinsic cause, develop in due course and by intrinsic self-direction the immediately exercisable capacity for characteristically human mental functions. . . .

. . . Since human beings are intrinsically valuable and deserving of full moral respect in virtue of what they are, it follows that they are intrinsically valuable from the point at which they come into being. Even in the embryonic stage of our lives, each of us was a human being and, as such worthy of concern and protection. Embryonic human beings . . . should be accorded the status of inviolability recognized for human beings in other developmental stages.[33]

This is perhaps the best defense of the claim that embryos are the equivalent of human persons, and I urge the reader to look at the entire statement of these scholars if she has the opportunity. Nonetheless, although this position is entitled to serious consideration, it is fundamentally flawed. This position is in tension with the accepted scientific understanding of embryonic development, is based on a controversial metaphysical position, conflicts with many of our commonly accepted moral judgments, and ultimately is unsupported by a credible theory of moral status. The position that embryos are entitled to the same moral status as human persons is untenable. Accordingly, there is no significant moral impediment to embryonic stem cell research.

Let us consider some of the flaws in this statement.

THE EARLY EMBRYO IS NOT AN INDIVIDUAL

As I have noted, until gastrulation, an embryo can divide into two or more parts, each of which, given appropriate conditions, might develop into separate human beings. This is the phenomenon known as "twinning" (although division into three or four separate parts is also possible). The phenomenon of twinning establishes that there is not one determinate individual from the moment of conception; adult humans are *not* numerically identical with a previously existing zygote or embryo. If that were true, then each of a pair of twins would be numerically identical with the *same* embryo. This is a logically incoherent position. If A and B are separate individuals, they cannot both be identical with a previously existing entity, C.

Many of those who contend that embryos are entitled to the same rights as human persons are aware of the twinning phenomenon but they discount its significance. First, they maintain that this phenomenon does not affect those embryos that do not separate. Second, even for those embryos that undergo twinning, they maintain that this process does not undermine the claim that there was at least one individual from the moment of conception. In the words of the 2002 majority report of the President's Council on Bioethics: "The fact that where 'John' alone once was there are now both 'John' and 'Jim' does not call into question the presence of 'John' at the outset."[34]

This reasoning is unpersuasive. If twinning does occur, and if "John" was there from the beginning and "Jim" originated later, this implies that at least some twins (and triplets, etc.) have different points of origin. This anomaly creates insuperable difficulties for a view that insists all human persons come into existence at the moment of conception. Are some twins not human?

Professor George, at least, is sufficiently concerned about the twinning phenomenon to have written an article recently (with Patrick Lee) devoted principally to discussing this issue. Their solution: "in twinning, either the first embryo dies and gives rise to two others, or the first embryo continues to live and a second embryo is generated upon the splitting of the first one."[35] In other words, if you do not like the implications of biology, create a new reality that provides you with the desired conclusions. There is no evidence that an embryo "dies" during the twinning process. No cell, no part of a cell, has ceased to live or function. George and Lee simply make this up. The same can be said for their unprecedented and unsupported assertion that humans can reproduce by asexual reproduction—which is what must happen if the dying embryo either "gives rise" to two others or another embryo splits off from the first. On no other occasion do humans reproduce asexually. (We could if we perfected reproductive cloning, but we're not there yet.) Lee and George insist humans must be able to do this, however, so they can preserve their metaphysics in light of the phenomenon of twinning. Essentially, their argument consists of nothing more than bare assertions and result-oriented reasoning.

It is also evident they have not thought through the implications of this new view. Effectively, they are saying that the twins born of the dead embryo—or the latter twin born by splitting off from the first embryo—are *not* children of the mother, but her grandchildren, if that. This has significant and far-reaching legal implications with respect to the inheritance of property, parental control, and so on. Presumably the first-born twin can successfully dispute the mother's right to raise the latter-born twin. After all, who's the parent? Any property left by the mother in her will to her "children" would not go to either twin if the first embryo has died. It would belong to the dead embryo, who presumably died intestate, and, regrettably, never had a chance to become acquainted with his/her/its children. Dad (or is it Mom?), we hardly knew ye. George and Lee's imaginative exercise in science fiction would be amusing were it not intended to be taken seriously as a basis for public policy that affects millions.

More important, the assertion that "John" is present from the outset—that is, there must be at least one individual present from the moment of conception—is nothing more than a dogmatic claim masquerading as scientific fact. There is no scientific evidence to establish the presence of a "John." What the science of embryonic development shows is that the early embryo consists of a grouping of cells with a genetic composition similar to the genetic composition of adult humans and that, after a period of time, under certain conditions, these cells begin to differentiate and organize themselves into a unified organism. Prior to gastrulation, there is no certainty that these cells will differentiate and organize, nor is there any certainty that these cells will become one, two, or more individuals. Prior to the controversies surrounding embryonic stem cell research, NIH actually established a panel of experts to study the status of the embryo. In the words of the Human Embryo Research Panel, the cells of an early embryo do not form part "of a coherent, organized, individual."[36] The phenomenon of twinning confirms that the early embryo is not a unified, organized, determinate individual. To insist otherwise is to maintain—without any supporting evidence—that there simply must be some occult organizing principles that we have not yet been able to detect. Effectively, the position that there simply must be a determinate individual from the moment of conception is a restatement of ancient ensoulment views in modern dress.

THE "POTENTIAL" OF THE EMBRYO DOES NOT MAKE IT A HUMAN PERSON

The fact that the early embryo is not an individual has obvious implications for the argument that the embryo is entitled to protection because it possesses the potential to develop capacities and properties characteristic of human persons. We cannot refer meaningfully to the potential of the embryo if it is not yet an individual.

However, even leaving the phenomenon of twinning aside, the argument from the potential of the embryo is not cogent for several reasons. The possibility that an embryo might develop into a human person does not obviate the fact that it has not yet acquired the capacities and properties of a person. An embryo is no more a human person than an acorn is an oak tree. Not only do embryos lack consciousness and awareness, but they do not have experiences of any kind, even of the most rudimentary sort. As already indicated, early embryos have not even undergone cell differentiation.

As indicated by the statement of Professors George and Gómez-Lobo, those who oppose embryonic research often try to minimize the gap between potential and actual possession of the characteristics of a human person by suggesting that the embryo's path of development is inevitable. They assert that from the moment a zygote is formed an embryo has the same genetic composition as the human person it will become and these genes provide it with the intrinsic capability of developing into that human person. Recall their statement that "if not prevented by some extrinsic cause, [the embryo will] develop in due course . . . by intrinsic self-direction." George and Gómez-Lobo have it backward. The embryo will not develop unless it is aided by extrinsic causes. Their argument completely overlooks the important role that extrinsic conditions play in embryonic and fetal development. An embryo in a petri dish is going nowhere. An embryo needs certain nutrients provided by the mother through the placenta in order to develop into a fetus and beyond. These nutrients regulate the epigenetic state of the embryo.[37] Those who claim full moral status for the embryo seem to regard gestation within a woman's uterus as an inconsequential and incidental detail. Obviously, it is not. The embryo must be provided with the appropriate conditions for development to occur. The embryo does not have the capability of expressing its "potential" on its own.[38]

That embryos do not have the intrinsic ability to develop into human persons on their own is demonstrated by the occurrence of teratomas and parasitic twins. Twins usually develop normally, given the right external conditions, but occasionally something goes off track, and one embryo becomes a tumor (teratoma) embedded inside the second twin. This teratoma often eventually possesses many of the indicia of a regular fetus, with hair, skin, teeth, and even hands, but it lacks organization. If possible, it is removed or destroyed. With parasitic twins, the removal may present more of a challenge, because the twin has developed further. In fact, portions of the parasitic twin may protrude outside the host's body. In other words, legs or arms of another being may project out of the host's abdomen, visible and waving about.[39]

Again, the messy reality of biology undercuts the convenient metaphysical categories imposed by those who regard the embryo as the equivalent of a human person. Unless the precisely appropriate external conditions and sequence of events are available to guide the embryo's development, the embryo will not become a human person.

Recognition of this fact has special relevance in the context of the debate over funding of stem cell research because of the current possible sources of embryos that may yield stem cells, namely, embryos from IVF procedures and embryos created from SCNT. In neither case are the embryos being removed from conditions that might permit their development. The spare embryos from IVF procedures have not been and will not be implanted in a uterus; instead, they will either be stored for an indefinite period or discarded. Therefore, they have no prospect of developing into a human person. Their potential is no more than a theoretical construct. (Obviously, the same would hold true for IVF embryos created with the specific intent to use them in research.)

The lack of any real potential to develop into a human person is even clearer in the case of embryos that might be created through SCNT. These embryos will be created with the specific intention of being used solely for research. Therefore, unless they are misappropriated by some pro-embryo activist and covertly implanted in a uterus, they have absolutely no chance of developing into a human person. It is misleading to speak of the potential of embryos to become human persons when the likelihood of such an event approaches zero.

Furthermore, the creation of embryos through SCNT shows that the argument from potential proves too much. Through SCNT, a somatic cell is allowed to express its potential to be transformed into an embryo that is latent in its genes but has been suppressed. If gene-based potential to develop into a human person is sufficient to provide an entity with full moral status, then *each* somatic cell in a human person's body has the same moral status as the person herself, because each cell has the potential to become a person, just as an embryo does. Contrary to what you might think, you are not an individual, but a collection of billions of individuals. And you can't even get a group discount when you buy tickets! The argument from potential leads to absurd conclusions, and, for that reason alone, should be rejected.[40]

THE UNACCEPTABLE CONSEQUENCES OF THE VIEW THAT THE EMBRYO IS A HUMAN PERSON

The conclusion that all the cells in a person's body possess the same moral rights as the person herself is just one of the unacceptable conclusions that follow from granting embryos the status of human persons. These unacceptable consequences demonstrate that granting full moral status to the embryo is not compatible with widely accepted moral norms and principles.

One important fact about embryonic development that is often overlooked is that between two-thirds and four-fifths of all embryos that are generated through standard sexual reproduction are spontaneously aborted.[41] If embryos have the same status as human persons, this is a horrible tragedy and public health crisis that requires immediate and sustained attention. *Up to 80 percent of humanity is at risk of dying suddenly!* Not only should we abandon stem cell research, but we should reallocate the vast majority of our research dollars from projects such as cancer research into programs to help prevent this staggering loss of human life. How can we have been so morally obtuse that we have failed to heed the silent cry of the hundreds of millions of embryos that "die" each year?

Interestingly, none of the opponents of embryonic stem cell research have called for research programs that might increase the odds of embryo survival. Their failure to address this issue is puzzling if the embryo deserves the same moral respect as human persons. Consider that great strides have been made in reducing infant mortality in the last century. Why do the opponents of embryonic stem cell research not demand that similar efforts be made to improve the survival rate of embryos?

Similarly, IVF, at least as currently practiced, would appear to be morally objectionable regardless of whether some embryos produced by this procedure are used in research. Those who utilize IVF intentionally create many embryos that they know will be discarded eventually. How can we accept a process that consigns entities that supposedly have the status of human persons to the rubbish bin? Hypotheticals can sometimes prove useful in testing our moral judgments. Consider what our moral reaction would be if we had a process that generated not embryos but infants at a developmental stage of about ten months. Would we regard this process as morally acceptable if the vast majority of infants so generated were thrown away? Presumably not. Indeed, many would find such a process repugnant. But if embryos have the same status as human persons, then a similarly repugnant result is produced by current IVF procedures.

Finally, it is worth noting that the focus of the current controversy over stem cell research is whether it should be federally funded, *not* whether it should be banned entirely (although there are some who have called for a ban). That we are even debating the wisdom of federal funding demonstrates that most of us—including the opponents of funding embryonic stem cell research—do not consider the embryo to have the same status as a human person. We do not debate the pros and cons of federal funding of research that would destroy adult humans.

Consideration of these implications of the position that embryos are entitled to the same rights as human persons demonstrates that this position cannot be reconciled with widely accepted moral norms and principles—including norms and principles accepted by opponents of funding embryonic stem cell research. Of course, this does not "prove" that this position is morally unsound. It merely establishes that their moral claims are incoherent. It is always open to advocates of this position to argue that accepted moral norms and principles, including their own, are in need of radical reform. Perhaps George, Lee, and Gómez-Lobo can start a *Save the Embryo Foundation* and organize fund-raisers for research into reducing the rate of "embryo mortality," or they can organize pickets of IVF clinics. However, to date no such call for action or for a revolution in our moral principles has issued from those who regard embryos as the equivalent of human persons.

THE FAILURE TO PROVIDE AN ADEQUATE THEORY OF MORAL STATUS

Some of the inconsistencies and astonishing implications of the view that the embryo has the status of a human person could be overlooked, at least temporarily, if the defenders of this view put forward a defensible theory of moral status. In my last objection, I note that those who insist that the embryo has the same moral status as a human person fail to articulate an adequate theory of moral status. In other words, they fail to identify which capacities or properties, intrinsic or relational, qualify an entity for moral respect. For example, is it rationality, the capacity for moral agency, sentience, social relationships or some combination of these that constitutes a necessary or sufficient condition for moral status?

The defenders of full moral status for the embryo typically rely solely on the assertion that "humanity" entitles one to full moral status and that the embryo is fully human in light of its genetic composition.[42] There are several problems with this claim, however.

To begin, if humanity is a necessary condition for moral status, then this would preclude granting moral status both to nonhuman animals and to extraterrestrials who exhibit capacities such as rationality or moral agency. Apparently, we not only have no obligation to let ET phone home, but we can chop him up and serve him for dinner. Without further argument, this would appear to be an arbitrary exclusion. Certainly, such entities appear to have interests (including the desire to be free from pain or distress) similar to human interests that are fostered or protected by our moral norms.

If humanity is a sufficient but not a necessary condition for moral status, then what is it about humanity that entitles one to moral status? No explanation is offered by those opposed to embryonic stem cell research other than the biological criterion of human genetic composition. However, unless a rationale is provided that explains *why* human genetic composition is so critical, then the insistence that genetic humanity is the key to moral status is mere question-begging.

Furthermore, insistence that genetic humanity is the key to moral status has disturbing implications. It is often overlooked that not all embryos, fetuses, or children produced by human parents have the same number of chromosomes. Humans typically have twenty-three pairs of (or forty-six) chromosomes. However, some embryos, fetuses, and children have extra chromosomes. Cells that have an irregular number of chromosomes are called aneuploid. Most embryos with aneuploid aberrations are spontaneously aborted, but some can survive. Down syndrome children, for example, have an extra chromosome 21. If genetic composition is what is critical for being a true human and enjoying full moral status, what do we say about Down syndrome children or children with other aberrant chromosomal composition? If they are entitled to full moral status, then genetic composition *cannot* be the sole determinant of moral status, but if it is not the sole determinant, what other factors are relevant and how would these other factors affect the status of embryos? To date, no proponent of the view that embryos are entitled to full moral status has satisfactorily answered these questions. But without answers to these questions, the claim that the embryo is equivalent to a human person cannot be adequately supported.

I posed this issue to Professor Gómez-Lobo in correspondence with him, asking him about his view that genetic composition is the critical criterion for moral status, and, if so, where does that leave children with Down syndrome or similar genetic abnormalities? His response essentially was that we need to accept as an empirical fact "that there are normal human genomes and abnormal ones." In his view this would not preclude treating Down syndrome children as humans. Indeed, he vigorously rejected any suggestion to that effect. He also candidly admitted there is a "normative" element to his claim about moral status. In other words, he maintains we *should* treat Down syndrome children as humans even if their genomes are abnormal.[43]

I do not quarrel with Gómez-Lobo's ultimate conclusion about the treatment that should be given children with Down syndrome, but his explanation does reveal gaps in his and Professor George's defense of the moral status of the embryo. At the end of the day, they are simply maintaining we *should* treat the embryo as though it were a human person, not that, as a factual matter, the embryo is a human person. It is perfectly permissible to advocate such a position, but for those not persuaded by their normative argument—and this appears to be the majority of the American people—there is no reason to accept their position as an expert's interpretation of scientific fact. To the contrary, for all the reasons already discussed, science tends to support the opposite conclusion; the embryo is not the equivalent of a human person.

Critics might say that it is incumbent upon those who defend embryonic stem cell research to provide their own theory of moral status. I accept the validity of this point—in part. I say in part because even if I cannot provide a thoroughly adequate theory of moral status, the fact that I have shown the other position (that is, the position that the embryo is a human person) cannot be sustained is sufficient to support my conclusions about the policy of funding embryonic stem cell research.

Although it is not possible to provide anything resembling a thorough and definitive argument for a theory of moral status within the confines of this chapter or even this book, I will provide the following outline of the elements of such a theory: I maintain that the scope of morality, which is a set of practices that ultimately relies on reason instead of force, should presumptively include all beings who are capable of reasoning and, therefore, capable of being influenced by moral norms. (This provides the underlying rationale for the intuition that rationality or moral agency is important for moral status.) Moreover, we should not lose sight of the fact that morality has objectives, one of which is to ensure the survival of the moral community, including oneself and one's loved ones, which, for most of us, includes our children. Our children embody our hopes and aspirations and assuming a moral community has a desire to survive for more than one generation, its children are the key to its survival. So babies who are wanted and intentionally gestated are entitled to the protection of our moral norms even when they are too young to be capable of reasoning. However, embryos that are designated for research use are, by definition, not entities that are, or have the potential to become, children and members of the moral community. Nor do they possess consciousness or rationality or any of the other characteristics that might otherwise entitle an entity to membership in the moral community. Accordingly, the fact that their genetic composition may be similar to members of the moral community does not, by itself, entitle these entities to the protections of our moral norms.

CONCLUSION: A REASONABLE POLICY ON STEM CELL RESEARCH

For the reasons I have set forth, the federal government should fund embryonic stem cell research. The primary objection to such funding is that the embryo is inviolable because it has the same status as a human person. While this position is maintained by a number of persons, including a few respected scholars, this position cannot withstand critical scrutiny. As illustrated by Tony Snow's futile search for the appropriate noun to describe the deliberate destruction of the embryo, even those who assert that an embryo is a human person still find themselves unable to equate the destruction of an embryo with the killing of a human person. Their internal emotional conflict and confusion mirrors the incoherence of their position on the embryo's status.

Perhaps because those who oppose embryonic stem cell research sense that they are not going to be successful in persuading most of us that an embryo is a human person, they have switched tactics of late. Beginning in 2007, the claim most often advanced by opponents of embryonic stem cell research is that it is not necessary to obtain stem cells from the inner cell mass of the blastocyst, and we should fund research on other types of stem cells or embryonic stem cells obtained through other means. Indeed, the White House issued a report in January 2007 that promoted so-called alternatives to embryonic stem cell research as it is currently practiced.[44] (The ideological character of the report was revealed by its title: *Advancing Stem Cell Science without Destroying Human Life*.) The recent successful development of iPS cells, unfortunately, has made the claim that we can forego use of embryonic stem cells appear more reasonable than it really is. In fact, many opponents of embryonic stem cell research have argued that the development of iPS cells vindicates Bush's decision not to fund such research. They maintain both that it is no longer necessary to use stem cells derived from embryos and that Bush's policy motivated researchers to search for "ethically neutral" alternatives.[45]

These claims are unfounded—and rash. The generation of iPS cells is a welcome development, but as already indicated, it is presently unknown whether iPS cells could be used in therapies. We simply do not know yet whether they hold the same promise as embryonic stem cells.

Much the same can be said for other proposed alternatives to embryonic stem cells. As previously noted, researchers have had limited success in inducing adult stem cells to differentiate into various types of tissue, and recent findings suggest that fetal stem cells found in amniotic fluid may be pluripotent. Research on these different types of stem cells definitely should be pursued. But to say that we should pursue these other methods *instead* of current methods of embryonic stem cell research is a misleading and reckless way to frame the issue. Research on iPS cells and adult and fetal stem cells can and should proceed in conjunction with research on embryonic stem cells. Focusing on iPS cells or fetal stem cells, while foregoing research on embryonic stem cells, is not the best way to advance our scientific knowledge. We need a comprehensive approach that will support research on all the various types of stem cells and stem cell equivalents in order to achieve the scientific breakthroughs necessary for a complete understanding of cell development.

Significantly, James Thomson, the researcher who first succeeded in cultivating stem cell lines from embryonic stem cells and who was one of the researchers who recently generated iPS cells, agrees with this analysis. In a recent commentary, Thomson, and coauthor Alan Leshner, have argued that "we don't yet know whether [iPS cells are] viable for treating human diseases," adding "we simply cannot invest all our hopes in a single approach. Federal funding is essential for both adult and embryonic stem cell research, even as promising alternatives are beginning to emerge."[46]

Thomson and Leshner also rejected the claim that the production of iPS cells vindicates Bush's opposition to embryonic stem cell research. They noted that far from establishing the wisdom of withholding federal funds from embryonic stem cell research, the recent production of iPS cells highlights some of the problems caused by this policy. Thomson and Leshner believe the production of iPS cells probably was delayed by Bush's opposition to embryonic stem cell research because "work by both the U.S. and Japanese teams that reprogrammed skin cells depended entirely on previous embryonic stem cell research."[47] Trying to conduct research on cell development while refusing to fund embryonic stem cell research is analogous to trying to determine the causes of global warming without funding research on greenhouse gas emissions. An adequate understanding of cell development—and of the therapies that may result from use of stem cells—would be delayed unnecessarily for years, if not completely impeded. These delays will come at the expense of millions of ill, injured, and suffering persons.

There is no moral justification for these delays. Hundreds of thousands of spare embryos from IVF procedures are available for research now. If they are not used for research, eventually they will be discarded. Saving embryos for the trash—that's the essence of Bush administration policy on embryonic stem cell research.

NOTES

1. See, for example, Gina Kolata, "Scientists Bypass Need for Embryo to Get Stem Cells," *New York Times*, November 21, 2007, p. A1; Rick Weiss, "Scientists See Potential in Amniotic Stem Cells," *Washington Post*, January 8, 2007, p. A1.

2. James A. Thomson et al., "Embryonic Stem Cell Lines Derived from Human Blastocysts," *Science* 282 (1998): 1145–47.

3. *Balanced Budget Downpayment Act*, U.S. Statutes at Large 110 (1996): 26.

4. National Institutes of Health, "Guidelines for Research Using Human Pluripotent Stem Cells," *Federal Register* 65 (August 25, 2000): 51976–81.

5. President George W. Bush, "President Discusses Stem Cell Research," August 9, 2001. Available at http://www.whitehouse.gov/news/releases/2001/08/20010809-2.html (accessed October 20, 2007).

6. Ibid.

7. Nicholas Wade, "Clinics Hold More Embryos Than Had Been Thought," *New York Times*, May 9, 2003, p. A24.

8. President George W. Bush, "President Discusses Stem Cell Research Policy," July 19, 2006. Available at http://www.whitehouse.gov/news/releases/2006/07/print/20060719-3.html (accessed October 20, 2007).

9. Press Briefing by Tony Snow, Office of the Press Secretary, the White House, July 18, 2006. Available at http://www.whitehouse.gov/news/releases/2006/07/20060718.html (accessed October 23, 2007). For the explanation of Snow's previous remarks, see Press Briefing by Tony Snow, Office of the Press Secretary, the White House, July 24, 2006. Available at http://www.whitehouse.gov/news/releases/2006/07/20060724-4.html (accessed October 23, 2007).

10. Michael A. Fletcher, "Bush Vetoes Stem Cell Research Legislation," *Washington Post*, June 21, 2007, p. A4.

11. National Institutes of Health, *Stem Cell' Basics*. Available at http//: stemcells.nih.gov/info/basics (accessed October 23, 2007.) See also President's Council on Bioethics, *Monitoring Stem Cell Research* (Washington, DC: President's Council on Bioethics, 2004), esp. pp. 8–11 and 111–28, available at http://www.bioethics.gov/reports (accessed October 23, 2007), and Irving L. Weissman, "Stem Cells—Scientific, Medical and Political Issues," *New England Journal of Medicine* 346 (2002): 1576–79.

12. National Institutes of Health, *Stem Cell Basics*.

13. Weissman, "Stem Cells—Scientific, Medical and Political Issues," p. 1576.

14. Heidi Ledford, "Doubts Raised over Stem-Cell Marker," *Nature* 449 (2007): 647.

15. National Institutes of Health, *Stem Cell Basics*.

16. Paolo De Coppi et al., "Isolation of Amniotic Stem Cell Lines with Potential for Therapy," *Nature Biotechnology* 25 (2007): 100–106.

17. Rick Weiss, "Scientists See Potential in Amniotic Stem Cells," *Washington Post*, January 8, 2007, p. A1 (quoting Richard Doerflinger of the US Conference of Catholic Bishops to the effect that the Catholic Church does not consider the use of fetal cells from amniotic fluid impermissible).

18. Irina Klimanskaya et al., "Human Embryonic Stem Cell Lines Developed from Single Blastomeres," *Nature* 444 (2006): 481–85.

19. Rick Weiss, "Lab Cites Stem Cell Advance," *Washington Post*, January 11, 2008, p. A4.

20. Ibid. See also Nancy F. O'Brien, "Pro-Life Official Dismisses New Stem-Cell Announcement as a Sham," *Catholic News Service*, August 28, 2006. Available at http://www.catholicnews.com/data/stories/cns/0604840.htm (accessed October 23, 2007).

21. Rick Weiss, "Mature Human Embryos Created from Adult Skin Cells," *Washington Post*, January 18, 2008, p. A1.

22. Insoo Hyun and Kyu Won Jung, "Human Research Cloning, Embryos, and Embryo-Like Artifacts," *Hastings Center Report* 36, no. 5 (2006): 34–41.

23. Paul R. McHugh, "Zygote and 'Clonote'—The Ethical Use of Embryonic Stem Cells," *New England Journal of Medicine* 351 (2004): 209–11.

24. President's Council on Bioethics, *Alternative Sources of Human Pluripotent Stem Cells* (Washington, DC: President's Council on Bioethics, 2005), esp. pp. 36–54.

25. Junying Yu et al., "Induced Pluripotent Stem Cell Lines Derived from Human Somatic Cells," *Science* 318 (2007): 1917–20.

26. See the discussion of this point in David DeGrazia, "Moral Status, Human Identity, and Early Embryos: A Critique of the President's Approach," *Journal of Law, Medicine and Ethics* 34 (2006): 49–57, and Bonnie Steinbock, "The Morality of Killing Human Embryos," *Journal of Law, Medicine and Ethics* 34 (2006): 26–34.

27. See the discussion in Lee M. Silver, *Challenging Nature: The Clash between Biotechnology and Spirituality* (New York: Harper Perennial, 2007), pp. 141–42.

28. Jong-Hoon Kim et al., "Dopamine Neurons Derived from Embryonic Stem Cell Function in an Animal Model of Parkinson's Disease," *Nature* 418 (2002): 50–56.

29. Deepa Deshpande et al., "Recovery from Paralysis in Adult Rats Using Embryonic Stem Cells," *Annals of Neurology* 60 (2006): 22–34.

30. For an argument defending this view, see my article "Slaves, Embryos, and Nonhuman Animals: Moral Status and the Limitations of Common Morality Theory," *Kennedy Institute of Ethics Journal* 15 (2005): 323–46.

31. Num. 31:17–18.

32. President's Council on Bioethics, *Monitoring Stem Cell Research*, p. 79.

33. President's Council on Bioethics, "Statement of Professor George (Joined by Dr. Gómez-Lobo)," *Human Cloning and Human Dignity* (New York: Public Affairs, 2002), pp. 294–301.

34. President's Council on Bioethics, *Human Cloning and Human Dignity*, p. 177.

35. Patrick Lee and Robert P. George, "The First Fourteen Days of Human Life," *New Atlantis*, Summer 2006, p. 65.

36. Human Embryo Research Panel, National Institutes of Health, *Report of the Human Embryo Research Panel* (Bethesda, MD: NIH, 1994), 1: 9.

37. Guoyao Wu et al., "Maternal Nutrition and Fetal Development," *Journal of Nutrition* 134 (2004): 2169–72.

38. In personal correspondence with me, Professor Gómez-Lobo explained he understood that an embryo could not be sustained without external help. However, he said what he meant by his statement in *Human Cloning and Human Dignity* is that the zygote after karyogamy (creation of the nucleus) has all the genetic material it will ever have and this genetic material will later direct the development of the embryo, fetus, infant, and adult. Alfonso Gómez-Lobo, e-mail letter to author, February 9, 2003. First, in the case of chimeras (fused

embryos) this is not factually correct. Genetic material will be added. Second, and more important, this does not obviate the fact that the embryo cannot develop on its own. Its genes may dictate how it will interact with its environment, but they do not control its environment or development.

39. Rowena Spencer, "Parasitic Conjoined Twins: External, Internal (Fetuses in Fetu and Teratomas), and Detached (Arcardiacs)," *Clinical Anatomy* 14 (2001): 428–44.

40. For similar reasons, the project of creating iPS cells through "dedifferentiation" poses moral problems for those who think an embryo has the status of a human person because of its potential. Recall that dedifferentiation involves reprogramming a somatic cell so it loses its specialization and becomes the functional equivalent of an embryonic stem cell. All one would have to do is move the cell back a bit further in its development and it could become an early embryo in a totipotent state. Thus, it has the intrinsic "potential" to become an embryo and then a person based on its genetic composition. This point has been overlooked or unacknowledged by those who argue that the development of iPS cells obviates the moral problems presented by use of stem cells derived from embryos.

41. President's Council on Bioethics, *Monitoring Stem Cell Research*, p. 88.

42. "Statement of Professor George (Joined by Dr. Gómez-Lobo)," pp. 294–301.

43. Alfonso Gómez-Lobo, e-mail letter to author, February 9, 2003.

44. Domestic Policy Council, the White House, *Advancing Stem Cell Science without Destroying Human Life*, January 9, 2007. Available at http://www.whitehouse.gov/stemcell (accessed October 27, 2007).

45. See, for example, Charles Krauthammer, "Stem Cell Vindication," *Washington Post*, November 30, 2007, p. A23.

46. Alan I. Leshner and James A. Thomson, "Standing in the Way of Stem Cell Research," *Washington Post*, December 3, 2007, p. A17.

47. Ibid.

Concluding Observations—And an Outline of a Theory of Moral Status

We have covered a lot of ground in this book. I thank you for your patience, and I hope our reasoning together about some selected issues in bioethics will encourage you to lend your voice to public policy discussions. Bioethics encompasses too many critical issues to be left to presumed experts, many of whom are self-designated. Obviously, we want input from scholars on these issues, but we also need input from the general public. How these public policy issues are resolved will affect all of us.

If you are interested in keeping up with developments in bioethics, there is a Web site that regularly posts news stories and commentary on bioethical issues. The site's address is: www.bioethics.net.*

There are obviously a number of other issues that could be discussed in a book such as this, but I believe I have already indicated how such issues should be addressed. We should start with the norms of the common morality, along with factual and moral beliefs widely accepted in our culture, and engage in a very rough weighing of the projected harms and benefits of following a proposed course of action. Then we should test our initial moral judgments by detailing and examining their consequences, trying to systematize them and their consequences in a set of general moral principles that can explain and account for these judgments. We must then attempt to render these judgments and principles consistent—a process that might require adjustment of both our judgments and our principles. Finally, it is also critically important that we understand the rationale for our norms.

*I have no affiliation with this site.

I want to expand on this last point for a bit, both because I think it is a consideration often overlooked and because I believe it is the key to resolving the troublesome question of moral status, that is, the issue of determining the beings to whom (or to which) we may have moral obligations and the extent of these obligations. One does not have to be a bioethicist to recognize that this is a knotty problem that appears to be at the core of some of our disputes, including disputes over abortion and animal rights and, to a lesser extent, the dispute over embryonic stem cell research.

Let me suggest that the way in which this problem is traditionally approached is mistaken. Philosophers and others who spend time addressing this issue typically start with the premise that there is some property or characteristic that is the key to having full moral status. Often, this property is thought to be some intellectual capacity, such as rationality or self-consciousness or moral agency, that is, an entity's ability to make moral judgments and take action consistent with those judgments. These seem superficially plausible. As I mentioned in the last chapter, because morality is a set of practices that ultimately relies on reason instead of force, it seems plausible to say that we should include within the scope of our moral norms all beings who are capable of reasoning and, therefore, capable of being influenced by moral norms.

But then we immediately run into problems, because young children and some cognitively deficient adults lack the capacity to be moral agents. This has resulted in some animal rights advocates deploying what has been unfortunately called the "argument from marginal cases." (I use "unfortunate" because its name seems to suggest that infants and the cognitively impaired are "marginal.") A number of philosophers, including Peter Singer, David DeGrazia, and Tom Regan, have used the argument from marginal cases to contend that animals deserve "equal consideration," that is, that we are morally obliged to give equal weight to the similar interests of humans and animals.[1] Distilled to its essential core, the argument from marginal cases is as follows: (1) There is no moral significance to membership within a particular species. (2) If we cannot locate moral significance in species membership, we must consider the capacities or properties of humans and animals. (3) If we use capacities or properties such as self-consciousness, rationality, moral agency, and the like to distinguish humans from animals, we also must conclude that young children and cognitively impaired humans are not deserving of the same treatment as other humans. (4) The conclusion that young children and cognitively impaired humans do not deserve equal consideration is unacceptable. (5) Therefore, assuming we wish to avoid this unacceptable conclusion, we are forced to concede that animals are owed equal consideration.

This argument seems to have some force; Peter Singer certainly believes it is an unanswerable argument.[2] Of course, for some philosophers and bioethicists the answer is that "humanity" is what gives us moral status—but Singer would argue that this is a question-begging response. Why should being human be necessary for moral status? That position seems to be a reflection of speciesism. What would happen if we encountered a race of intelligent extraterrestrials? Would we say we have no obligations to them because they are not human?

Let me suggest that lost in all this back-and-forth is another question that needs to be addressed by philosophers on all sides of this issue (as well as anyone else interested in this issue): Why should we think that the mere possession of *any* capacity or property is the key to moral status?

A different approach to the question of moral status in general, and of animals in particular, is both needed and warranted. Instead of focusing on intrinsic capacities or properties, we should consider the goals we seek to accomplish through our moral norms and how the status granted a particular entity may advance or hinder these goals.

As I have stated previously in this book, I view morality as a practical enterprise. Moral norms serve certain purposes. Moral norms, such as the injunctions not to kill others or deprive them of certain goods, or to provide some measure of assistance to those in distress, enable individuals within groups that accept such norms to achieve a less painful, more desirable existence, by, among other things, helping to provide more security to these individuals, ameliorating harmful conditions, and, in general, facilitating cooperation in achieving shared, complementary, or individual goals. Moral norms enable us to live together, with the benefits derived from communal activity, while maintaining some discretion over our personal activities.

As is true with any attempt to influence conduct, moral norms have not always been effective. But they have been sufficiently successful. Indeed, the success of moral norms in achieving their objectives provides at least part of the explanation for their acceptance. That moral norms help achieve certain commonly desired objectives (security, assurance of support in confronting hazards and threats, stability in possessions and the ability to obtain goods that one cannot produce alone) also provides at least part of the explanation for the prevalence of certain core norms throughout history and across all cultures. There has never been a human community that did not have some restrictions on killing, maiming, and stealing from recognized members of the community.[3]

However, if morality is a practical enterprise, we may need to limit or withhold moral rights from particular sets of entities. As a foundational premise for this pragmatic approach, we should adopt the following proposition: We should not grant equal moral status to certain entities if doing so would tend to undermine our ability to achieve the objectives of morality. In other words, morality should not be self-defeating. This proposition may appear to be fairly innocuous—to the point of being uninformative and useless—but it is actually a proposition whose acceptance would reframe the debates about moral status. This proposition implies that the scope of morality is contingent on prevailing conditions, including the ability of individuals to coordinate their conduct with each other and to inculcate norms that will further the objectives of morality. Not only do animals and marginal humans lack an inherent, unquestionable entitlement to moral status, but "normal" adult humans lack this entitlement as well.

I recognize that some may find this contention deeply disquieting. Nonetheless, at the end of the day, if we abandon the belief that moral norms *must* apply equally to all humans, or all humans who possess certain capacities or properties, we will gain a better understanding of moral status and of the considerations that are relevant to resolving some of our disputes about moral status, including the status of animals.

In addressing the relationship of moral status to the objectives of morality, we should consider circumstances other than our currently prevailing conditions. We should consider a situation in which the following conditions prevail among groups of humans: little or no interaction between these groups other than conflict; an inability to understand many of the motivations and intentions of humans in other groups; no trust between groups. In other words, we should consider a situation similar to that in which most humans lived for millennia—at least according to many anthropologists.[4]

One reason it is necessary to consider this set of circumstances is to test whether our intuitions about moral status, in particular, the received views about the consideration due other humans, implicitly rely on our current conditions, that is, conditions in which we can advance the objectives of morality by inclusion of all humans within the scope of our norms. For a few centuries now we have had significant interaction among humans across the globe, with the interdependence of human groups generally becoming even more pronounced in recent times. Advances in technology place us in immediate and direct contact with virtually any corner of the globe. More important, repeated instances of individuals from many different cultures working together successfully—an occurrence that takes place millions of times each day—confirm that we can work cooperatively with most humans. This is a critical development, and one too often overlooked by many philosophers. The building of trust is a cumulative process and some level of trust between individuals is essential if moral norms are to fulfill their function. There is now sufficient implicit trust among most humans that one can safely say that the default mode of interaction between humans throughout this world is one in which moral norms govern our conduct. (By saying it is the default mode, I mean we are expected to treat other humans with equal moral consideration, not that we always act as though others had comparable moral status.) Under these circumstances, the view that all humans deserve equal moral status comes easily to us.

This was not true in earlier times. The view that our moral norms should encompass all, or nearly all, humans has gained widespread acceptance in the last few centuries, but for most of human history, the scope of our moral norms has been relatively circumscribed. Anthropologist Lawrence Keeley, in his seminal work on primitive warfare, has noted that not only was warfare incessant among primitive peoples, but it was also merciless.[5] Few prisoners were taken and those that were suffered horribly. In the previous chapter, I provided a vivid illustration of the prevailing attitude in earlier cultures by describing the conduct of Moses, someone often held out as one of our great moral teachers, who was an unabashed advocate of the ruthless killing and dispossession of humans who did not share his religious views. But Moses was not unique in this respect. For most of our existence as a species, moral norms were not considered to apply (or apply with equal force) to humans outside of one's group. Instead, moral obligations were limited to members of one's tribe or clan. Those outside this group had very limited rights, or no rights at all. In fact, most of our ancestors did not regard themselves as members of a global human society. Instead, they were members of distinct human groups, and being born a Canaanite most decidedly did not make one a member of Hebrew society. The notion that we have similar moral obligations to all humans, regardless of their particular origins, has gained widespread acceptance only in recent times.

It is tempting to dismiss the moral attitudes of prior generations as benighted and rife with unjustifiable prejudice. But the conditions under which most of our ancestors lived were not the conditions that prevail currently. Warfare was constant among primitive peoples and hostility between members of different communities was prevalent until recent times. Enmity and violence were the default modes of dealing with those outside of one's community because they appeared to be the most successful ways of dealing with outsiders, if by "successful" we understand an approach that is likely to ensure the survival, if not the flourishing, of oneself and those close to one. For moral norms to displace violence and force as the preferred method of interaction between groups of humans, some background conditions are necessary. Among them are a level of contact that creates the possibility for significant communication, a desire to maintain this level of contact, an ability to understand the motivations and intentions of others and, most critically, an ability to trust others. When others are deeply suspicious about one's intentions, this usually does not bode well for the treatment that one can expect to receive at their hands. There are too many transaction costs involved (including, potentially, the loss of their lives) for others to take the time to determine whether one's intentions are peaceful.

But isn't impartiality—the notion that I should treat others the way I want to be treated—the crux of morality and, if so, must not our moral norms include all humans, or at least all those capable of being moral agents? I grant that moral norms must incorporate some element of impartiality, but impartiality toward whom? I can treat members of my community impartially, while having little regard for "barbarians" and other outsiders. Impartiality cannot logically compel me to adopt a perspective that includes other societies within the scope of my community's moral norms if relations between our societies do not satisfy the minimum conditions necessary for the successful implementation of these norms. No matter how impartial our point of view, there is no moral imperative to accord comparable status both to those willing to work with me and to those who not only resist cooperation but who stubbornly persist in efforts to harm me and other members of my community. Impartiality cannot render us insensible to patterns of behavior that make our moral norms ineffective.

By this point in my discussion, I'm sure some readers are reacting negatively. First, let me reassure you I have no wish to return to the past. A world in which there was insurmountable hostility between groups of humans is a world well lost. Fortunately, the conditions under which our ancestors lived have vanished—never to return, one hopes.

But it is precisely because the conditions in our current world are so different that I believe these conditions supply implied and unacknowledged premises for the claim that the capability to make moral judgments by itself automatically entitles one to moral status. We may encounter the occasional person who appears unrestrained by many moral norms—in other words, the self-centered, amoral individual who only cares about "Number One"—but most of us never encounter someone who *is* restrained by moral norms, only insofar as his conduct is directed toward fellow members of his tribe. As a result, we simply overlook some of the conditions required for extending moral status to other entities. We ignore the material conditions necessary for moral status and focus instead on intrinsic properties. That, in turn, leads to unyielding and unnecessary puzzles and disputes about moral status, such as the dispute over the argument from marginal cases.

Let me try to drive home my point by asking the reader to consider a hypothetical. I referred to extraterrestrials earlier and how the fact that they were not human would not seem to exclude them from the scope of our moral norms. True. It's not a question of their being human or having the capacity for moral agency. It's a question of whether bringing such beings within the scope of our moral community would be self-defeating. Imagine a race of beings, let's call them the Spores, that attacks humanity mercilessly and continues attacking for centuries. Moreover, efforts to negotiate with the Spores prove uniformly unavailing. Emissaries from humanity who do not take appropriate precautions are killed and consumed. It is not even clear whether we humans are successful in communicating with the Spores because they do not acknowledge our efforts. Nonetheless, we attribute rationality to the Spores because their own conduct seems coordinated and their attacks carefully planned. Spores also appear to value other Spores because efforts are made to rescue captured Spores, Spores share nutrition (humans) with other Spores, and so on. One particular problem in dealing with the Spores is that they appear to engage in asexual reproduction, as isolated, captured Spores soon produce similar beings. Humanity, therefore, decides to kill all captured Spores. In short, we treat all Spores as outside the scope of our moral norms. Can anyone plausibly maintain that our treatment of the Spores is morally wrong?

This hypothetical can be embellished to the extent necessary, but the foregoing seems to provide enough detail to make the point—the point being that there can be circumstances in which we would be justified in refusing to treat some moral agents with equal consideration. Welcoming the Spores into our moral community not only would fail to further the objectives of our moral norms, but would thwart them.

So what does all this have to do with animals or children? One virtue of my approach to grounding moral status, at least in part, on prevailing conditions, is that it avoids any question-begging privilege for members of the human species. Neither humans with normal capacities nor humans lacking some capacities are necessarily members of the moral community. To this extent, my approach incorporates one of the insights of the argument from marginal cases, as well as other arguments in favor of moral obligations toward animals, which is that we should not merely assume that we have more extensive moral duties toward humans than we have toward animals. We need to justify any differential treatment.

At least theoretically then, animals could be entitled to equal moral consideration. But some animals (vermin, etc.) have interests inimical to ours. Many animals do not have, nor are they likely to have, the ability to live in harmony with us. Including them within the scope of the moral community would work against the achievement of morality's objectives.

Furthermore, most animals lack the ability to coordinate their conduct with our conduct, at least to any significant extent. This is an important impediment to integration within the moral community. Note that in making this observation, I am not insisting that the moral community be limited to those who can engage in reciprocal behavior, strictly understood. (One of the faults of some contract theories of morality is they seem to imply that moral status is based strictly on the capacity for reciprocal behavior.) Coordination does not always involve reciprocity. We can coordinate our actions with others simply by reaching an understanding that allows us to pursue our separate goals while refraining from conduct harmful to each other. But most animals cannot coordinate with us even to this limited extent. An illustration of this inability to coordinate their conduct with our conduct is the behavior of animals susceptible to infection from avian flu. In an ideal world, we would inform ducks and chickens of the danger and ask them to take appropriate precautions. Unfortunately, we cannot tell ducks and chickens what measures to take to avoid spread of this disease among them or us, so we kill them. The millions of birds destroyed without compunction because of the threat of avian flu bear mute witness to some of the obstacles to including animals within the scope of our norms.

Additional barriers to integration of most animals within the moral community are relative infrequency of contact and difficulties in communication and understanding that go beyond the inability to coordinate conduct. It is not because "they lack syntax, therefore we can eat them," as some animal rights advocates sarcastically observe. It is that their inability to communicate reliably with us precludes the development of extensive shared interests as well as a thorough understanding of their particular interests and desires. Granted there are some remaining formidable cultural barriers to understanding among humans, but these pale in significance to our incomprehension of the desires and interests of animals. For example, even with respect to as fundamental an issue as life versus death, it is not clear to us the extent to which animals are harmed by death. If we regard death as a harm principally because it thwarts an individual from fulfilling desires and projects, then, at least for many animals, death does not appear to be a significant harm, since they lack future-oriented desires and projects. For other animals it may be a significant harm, but, at least arguably, it is not as great a harm as it is for humans.

There is another obstacle as well, which is related to our limited ability to inculcate and apply moral norms. Advocates of equal consideration quite properly object vigorously to those who mischaracterize their views by suggesting that equal consideration requires equal treatment. Equal consideration does *not* mandate equal treatment. We are not obliged to give Fido the right to vote. But the fact that equal consideration not only allows for but presumably requires different treatment for different types of animals actually underscores one of the major problems in extending equal moral status to animals. The need to weigh the differing interests of innumerable species of animals presents a major difficulty to integrating animals within the moral community. As I discussed in chapter 2, moral norms must have a certain level of simplicity and generality if they are to be inculcated and applied successfully. There are nontrivial limits on the specificity of moral norms if we hope to inculcate these norms in our children so that they will become morally virtuous, that is, engage in acceptable behavior routinely and spontaneously. (Recall that this is one distinction between legal codes and moral norms. No one can know all the hundreds of thousands of legal regulations that govern conduct in contemporary societies. But that's fine, because no one is expected to conform spontaneously to these regulations. One learns the provisions of the tax code as needed.) It would be impossible to transmit, learn, and apply a set of moral norms that specified exactly what one should do in every possible situation, even if one could be devised. But the list of questions we would have to answer before we are to give animals equal moral consideration would fill volumes, and we would need an algorithm of staggering complexity in order to apply our moral norms successfully.

Furthermore, figuring out how to incorporate all animals as full members of the moral community is just the first step. Somehow we would have to train nearly all humans to adopt and follow the norms that give equal moral consideration to animals. This requires habituating adults in all societies to the complex norms that follow from granting equal moral status to animals. If moral norms are to be enforced, they require the approbation of nearly all moral agents. Such an immense project of moral reform is daunting, to say the least. How is such a project even to begin? Where is the disposition to undertake such a wholesale revision of our morality? Are humans even capable of adopting such an enormously complex moral code? These questions need to be answered before we can decide whether equal consideration is a plausible practical guide to ethical conduct; but these questions have not even been posed, let alone answered, by most animal rights advocates. As this chapter has indicated, it has taken humans millennia to reach the point where we can say there is sufficient understanding between different human societies that we can speak meaningfully of a global human community. To expect humans to achieve a comparable understanding of innumerable species of animals within a short period of time manifests much wishful thinking.

One critical point that is overlooked by proponents of equal consideration is that morality is fundamentally a means of adjusting our behavior so we can live together. We do not, for the most part, live with animals. There is no need to adjust our behavior to live with them, no particular motivation to adjust our behavior to live with them, and no reliable understanding of what it would mean to take their interests into account even if were we disposed to do so.

Of course, we do have an understanding of most animals to this extent: pain is a harm for them as it is for us. A moral norm prohibiting the infliction of gratuitous pain is a norm that probably has almost as much relevance to many animals as it does to humans. It is a norm that is easily understood and easily applied. Humans do not need special, prolonged training in a norm forbidding cruelty to know when to apply it. Furthermore, the development of a disposition to avoid cruelty to animals flows naturally from a disposition to avoid cruelty to fellow humans. Many of us respond instinctively to the cry of a kitten just as we respond instinctively to the cry of a baby. It is no accident that the moral norm regulating behavior toward animals that is most widely accepted by humans is the norm forbidding cruelty.

This is an appropriate juncture at which to note that entities can have degrees of moral status. The relations we have with animals are different than the relations we have with humans, and their capacities are (usually) different than our capacities. But few seriously dispute that many animals can experience pain, pleasure, and emotions. Even if animals are not full members of the moral community, this does not imply that we have no moral duties to them whatsoever. Arguably, if we scrupulously avoided all contact with animals and just let them live on their own, we might have no moral obligations to them. But, of course, that is not our situation. Among other things, we raise them to feed ourselves, train them to undertake tasks for us, and invite them into our homes as companions. Under these circumstances, we have made some of them part of our society and have assumed some responsibility for them. Animals do not deserve equal consideration, but they have a degree of moral considerability and our conduct toward them should be commensurate with their status. As indicated, for many animals this may not imply anything beyond a duty to refrain from cruelty, but as far as animals are concerned this may be the most important duty that we could have toward them. Other animals, especially companion animals, at least arguably have a higher degree of moral considerability.

At this stage, the animal rights advocate might suggest that, regardless of the validity of my thesis that social conditions are important for determining moral status, I have done nothing to refute the contention that so-called marginal humans are not distinguishable from animals in terms of their moral status. Humans with limited cognitive capacities have difficulty coordinating their conduct with us as much as animals do.

That may be the case, but difficulty in coordinating conduct is merely one of the obstacles faced by animals. Recall that the interests of some animals are hostile to those of humans. Moreover, with respect to those that have interests that are not hostile, we have great difficulty in understanding their interests and it is doubtful whether we could devise and inculcate moral norms that would protect and advance their interests. Not so with humans with limited capacities. They lack some of the interests of humans with standard capacities, but their remaining interests are more readily understood than the interests of animals and they fit comfortably within the scope of existing moral norms. Indeed, the project of carving out exceptions to our moral norms for humans with limited capacities would be a difficult task and one that would have to be undertaken with caution—even in instances where there may be good reason to recognize an exception. It is not a question of there being no sharp boundaries between humans with normal and humans with limited capacities. It is a question of how to incorporate these distinctions within a set of norms that can be successfully inculcated and applied. Our moral norms form part of an interlocking web of attitudes, beliefs, and dispositions, and norms that have been in place for extended periods of time cannot be substantially modified through the creation of multiple exceptions without threatening to make our moral code unduly complex, uncertain, and fragile. A moral system that requires use of a flow chart is unlikely to generate the requisite set of moral dispositions. Thus, for sound cultural and psychological reasons, we have grounds for treating humans with limited capacities differently than we treat animals.

But I have saved the most important and obvious distinction for last. The vast majority of "marginal" humans are infants who not only will develop standard capacities and properties within a relatively short period of time but also have strong social and emotional bonds with the other members of the moral community. Indeed, most of us have such strong emotional bonds with our children that saying we have an "obligation" to protect and nurture them somehow diminishes the significance of those bonds. Furthermore, to assert that extending the protection of moral norms to infants serves the objectives of morality would be an understatement. One of the principal objectives of morality is to provide some measure of security not only to ourselves but to our loved ones, and that includes our children. Our children embody our hopes and aspirations, and assuming a moral community has a desire to survive for more than one generation, its children are the key to its survival. Human children ensure our community has a future; animals do not. Only theories of moral status that have a misguided focus on capacities and properties could ignore this fundamental point.

The foregoing sketch of how issues of moral status should be addressed by focusing on the objectives of morality is just that—an incomplete picture with many details left to be filled in. Nonetheless, even this outline should be sufficient to show the key deficiencies in the prevailing method of addressing issues of moral status by focusing on intrinsic properties. As for filling in the picture with the necessary details—well, I have to leave something for my next book.

NOTES

1. See David DeGrazia, *Taking Animals Seriously* (Cambridge: Cambridge University Press, 1996); Tom Regan, *The Case for Animal Rights* (Berkeley: University of California Press, 1983); Peter Singer, *Animal Liberation* (New York: New York Review Books, 1975).

2. Peter Singer, *In Defense of Animals: The Second Wave* (Oxford: Blackwell, 2006), pp. 3–4.

3. See Sissela Bok, *Common Values* (Columbia: University of Missouri Press, 1995), esp. pp. 10–28. Turnbull's study of the Ik is often cited as evidence contrary to this proposition, but Ik society had already disintegrated at the time of Turnbull's study. See Colin Turnbull, *The Mountain People* (New York: Simon & Schuster, 1972).

4. See, for example, Lawrence H. Keely, *War before Civilization* (New York: Oxford University Press, 1996); Robert L. Caneiro, "Chiefdom-Level Warfare as Exemplified in Fiji and the Cauca Valley," in *The Anthropology of War*, ed. Jonathan Haas (Cambridge: Cambridge University Press, 1990), pp. 190–211.

5. Keely, *War before Civilization*, pp. 32–33, 83–106.

Bibliography

Adams, Ted. D., Richard E. Gress, Sherman C. Smith, R. Chad Halverson, Steven C. Simper, Wayne D. Rosamond, Michael J. LaMonte, Antoinette M. Stroup, and Steven C. Hunt. "Long-Term Mortality after Gastric Bypass Surgery." *New England Journal of Medicine* 347 (1998): 753–61.

Albert, Steven M., J. G. Rabkin, M. L. Del Bene, T. Tider, I. O'Sullivan, L. P. Rowland, and H. Mitsumoto. "Wish to Die in End-Stage ALS." *Neurology* 65 (2005): 68–74.

Alloy, Lauren B., and Lyn Y. Abramson. "Depressive Realism: Four Theoretical Perspectives." In *Cognitive Processes in Depression*, 223–65. Edited by Lauren B. Alloy. New York: Guilford, 1998.

American Psychiatric Association. *Diagnostic and Statistical Manual of Mental Disorders*, 4th ed. rev. Washington, DC: American Psychiatric Association, 1994.

Animal and Plant Health Inspection Service, Department of Agriculture. "Bayer Crop Science: Extension of Determination of Nonregulated Status to Rice Genetically Engineered for Glusofinate Herbicide Tolerance." *Federal Register* 71 (2006): 70360–62.

Annas, George J. "The Man on the Moon, Immortality, and Other Millennial Myths: The Prospects and Perils of Human Genetic Engineering." *Emory Law Journal* 49 (2000): 753–82.

———. "Physician-Assisted Suicide—Michigan's Temporary Solution." *New England Journal of Medicine* 328 (1993): 1573–76.

Annas, George J., Lori B. Andrews, and Rosario M. Isasi. "Protecting the Endangered Human: Toward an International Treaty Prohibiting Cloning and Inheritable Alterations." *American Journal of Law and Medicine* 28 (2002): 151–78.

St. Augustine. *The City of God.* Translated by Henry Bettenson. New York: Penguin Classics, 2003.

Axelrad, Jacob. *Patrick Henry: The Voice of Freedom.* New York: Random House, 1947.

Bailey, Ronald. *Liberation Biology: The Scientific and Moral Case for the Biotech Revolution.* Amherst, NY: Prometheus Books, 2005.

Battista, Judy. "Sideline Spying: N.F.L. Punishes Patriots' Taping." *New York Times*, September 14, 2007, p. A1.

Beauchamp, Tom L. "A Defense of the Common Morality." *Kennedy Institute of Ethics Journal* 13 (2003): 259–74.

Beauchamp, Tom L., and James F. Childress. *Principles of Biomedical Ethics*. 5th ed. New York: Oxford University Press, 2001.

Beauchamp, Tom L., and Arnold I. Davidson. "The Definition of Euthanasia." In *Moral Problems in Medicine*, 2nd ed. Edited by Samuel Gorovitz, John M. O'Connor, Ruth Macklin, Andrew Jameton, and Eugene V. Perrin, 446–58. Englewood Cliffs, NJ: Prentice Hall, 1983.

Belkin, Lisa. "Doctor Tell's of First Death Using His Suicide Device." *New York Times*, June 6, 1990, p. A1.

Bok, Sissela. *Common Values*. Columbia: University of Missouri Press, 1995.

Boyle, Joseph. "Sanctity of Life and Suicide: Tensions and Development within Common Morality." In *Suicide and Euthanasia*. Edited by Baruch A. Brody, 221–50. Dordrecht, Netherlands: Kluwer, 1989.

Breitbart, William, Barry Rosenfeld, Hayley Pessin, Monique Kaim, Julie Funesti-Esch, Michele Galietta, Christian J. Nelson, and Robert Brescia. "Depression, Hopelessness, and Desire for Hastened Death in Terminally Ill Patients with Cancer." *Journal of the American Medical Association* 284 (2000): 2907–11.

Buchanan, Allen, Dan W. Brock, Norman Daniels, and Daniel Wikler. *From Chance to Choice*. New York: Cambridge University Press, 2001.

Bush, George W. "President Discusses Stem Cell Research." August 9, 2001. http://www.whitehouse.gov/news/releases/2001/08/20010809-2.html (accessed October 20, 2007).

———. "President Discusses Stem Cell Research Policy." July 19, 2006. http://www.whitehouse.gov/news/releases/2006/07/print/20060719-3.html (accessed October 20, 2007).

Callahan, Daniel. *The Troubled Dream of Life: Living with Mortality*. New York: Simon & Schuster, 1993.

Caneiro, Robert L. "Chiefdom-Level Warfare as Exemplified in Fiji and the Cauca Valley." In *The Anthropology of War*, edited by Jonathan Haas, 190–211. Cambridge: Cambridge University Press, 1990.

Carpenter, Janet E., and Leonard P. Gianessi. *Agricultural Biotechnology: Updated Benefit Estimates*. Washington, DC: National Center for Food and Agricultural Policy, 2001. http://ncfap.org/reports/biotech/updatedbenefits.pdf (accessed September 30, 2007).

Centers for Disease Control and Prevention. *Foodborne Illness*. Washington, DC: Centers for Disease Control and Prevention, Department of Health and Human Services, 2005. http://www.cdc.gov/ncidod/dbmd/diseaseinfo/foodborne_illness_FAQ.pdf (accessed August 26, 2007).

———. *What Would Happen If We Stopped Vaccinations?* Washington, DC: Centers for Disease Control and Prevention, Department of Health and Human Services, 2007. http://www.cdc.gov/vaccines/vac-gen/whatifstop.htm (accessed September 19, 2007).

Childress, James F. "Ethics and the Allocation of Organs for Transplantation." *Kennedy Institute of Ethics Journal* 6 (1996): 397–401.

Child Welfare Information Gateway. *Child Abuse and Neglect Fatalities: Statistics and Interventions*. Washington, DC: Government Printing Office, 2006.

Cohen, Joshua T., and Peter J. Neumann. "What's More Dangerous, Your Aspirin or Your Car? Thinking Rationally about Drug Risks (and Benefits)." *Health Affairs* 26 (2007): 636–46.

Committee on Identifying and Assessing Unintended Effects of Genetically Engineered Foods on Human Health, National Research Council. *Safety of Genetically Engineered Foods: Approaches to Assessing Unintended Health Effects*. Washington, DC: National Academies Press, 2004.

Conko, Gregory. "The Benefits of Biotech." *Regulation* 26 (2003): 20–25. http://www.cato.org/pubs/regulation/regv26n1/v26n1-4.pdf (accessed August 19, 2007).

Council of Europe. "Convention for the Protection of Human Rights and Dignity of the Human Being with Regard to the Application of Biology and Medicine: Convention on Human Rights and Biomedicine." *ETS*, no. 164 (Oviedo, 1997), chap. 4, art. 13. http://conventions.coe.int.Treaty.en/Treaties/Word/164.doc (accessed October 18, 2007).

Curlin, Farr A. "Caution: Conscience Is the Limb on Which Medical Ethics Sits." *American Journal of Bioethics* 7 (2007): 30–32.

Daniels, Norman. *Just Health Care*. New York: Cambridge University Press, 1985.

De Coppi, Paolo, Georg Bartsch Jr., M. Minhaj Siddiqui, Tao Xu, Cesar C. Santos, Laura Perin, Gustavo Mostoslavsky, Angéline C. Serre, Evan Y. Snyder, James J. Yoo, Mark E. Furth, Shay Soker, and Anthony Atala. "Isolation of Amniotic Stem Cell Lines with Potential for Therapy." *Nature Biotechnology* 25 (2007): 100–106.

DeGrazia, David. "Moral Status, Human Identity, and Early Embryos: A Critique of the President's Approach." *Journal of Law, Medicine and Ethics* 34 (2006): 49–57.

———. *Taking Animals Seriously: Mental Life and Moral Status*. New York: Cambridge University Press, 1996.

Deshpande, Deepa, Yun-Sook Kim, Tara Martinez, Jessica Carmen, Sonny Dike, Irina Shats, Lee Rubin, Jennifer Drummond, Chitra Krishnan, Ahmet Hoke, Nicholas Maragakis, Jeremy Shefner, Jeffery Rothstein, and Douglas Kerr. "Recovery from Paralysis in Adult Rats Using Embryonic Stem Cells." *Annals of Neurology* 60 (2006): 22–34.

Diamond, Jared. *Guns, Germs, and Steel*. New York: W. W. Norton, 1999.

Doerflinger, Richard. "Assisted Suicide: Pro-Choice or Anti-Life?" *Hastings Center Report* 19 (1989): S16–19.

Domestic Policy Council. *Advancing Stem Cell Science without Destroying Human Life*. January 9, 2007. http://www.whitehouse.gov/stemcell (accessed October 27, 2007).

Emanuel, Ezekiel, E. R. Daniels, D. L. Fairclough, and B. R. Clarridge. "Euthanasia and Physician-Assisted Suicide: Attitudes and Experiences of Oncology Patients, Oncologists, and the Public." *Lancet* 347 (1996): 1805–10.

Engelhardt, H. Tristram, Jr. "The Disease of Masturbation: Values and Concepts of Disease." *Bulletin of the History of Medicine* 48 (1974): 234–48.

Federal Bureau of Investigation. *Expanded Homicide Data, Table 9. Crime in the United States 2005*. Washington, DC: Federal Bureau of Investigation, 2006. http://www.fbi.gov/ucr/05cius/offenses/expanded_information/data/shrtable_09.html (accessed July 14, 2007).

Federoff, Nina V. "Agriculture: Prehistoric GM Corn." *Science* 302 (2003): 1158–59.

Feinberg, Joel. *Harm to Self*. New York: Oxford University Press, 1986.

Fine, Gary Alan. "Mercantile Legends and the World Economy: Dangerous Imports from the Third World." *Western Folklore* 48 (1989): 153–62.

Fletcher, Michael A. "Bush Vetoes Stem Cell Research Legislation." *Washington Post*, June 21, 2007, p. A4.

Food and Drug Administration. "Statement of Policy: Foods Derived from New Plant Varieties." *Federal Register* 57 (1992): 22984–23005.

Fromartz, Samuel. *Organic, Inc.: Natural Foods and How They Grew*. New York: Harcourt, 2006.

Fukuyama, Francis. *Our Posthuman Future: Consequences of the Biotechnology Revolution.* New York: Farrar, Straus and Giroux, 2002.

Ganzini, Linda, Thomasz M. Beer, Matthew Brouns, Motomi Mori, and Yi-ching Hieh. "Interest in Physician-Assisted Suicide among Oregon Cancer Patients." *Journal of Clinical Ethics* 17 (2006): 27–38.

Ganzini, Linda, Theresa A. Harvath, Ann Jackson, Elizabeth R. Goy, Lois L. Miller, and Molly A. Delorit. "Experiences of Oregon Nurses and Social Workers with Hospice Patients Who Requested Assistance with Suicide." *New England Journal of Medicine* 347 (2002): 582–88.

Ganzini, Linda, Heidi D. Nelson, Terri A. Schmidt, Dale F. Kraemer, Molly A. Delorit, and Melinda A. Lee. "Physicians' Experiences with the Oregon Death with Dignity Act." *New England Journal of Medicine* 342 (2000): 557–63.

Garcia, Jorge L. A. "Better Off Dead?" *APA Newsletter on Philosophy and Medicine* 92 (1993): 85–88.

Gert, Bernard. *Morality: Its Nature and Justification.* Rev. ed. New York: Oxford University Press, 2005.

Gorsuch, Neil M. *The Future of Assisted Suicide and Euthanasia.* Princeton, NJ: Princeton University Press, 2006.

Green, Ronald M. *The Human Embryo Research Debates.* New York: Oxford University Press, 2002.

Gunther, Marc. "Attack of the Mutant Rice." *Fortune*, July 9, 2007.

Guttmacher Institute. *State Policies in Brief: Refusing to Provide Health Services.* New York: Guttmacher Institute, 2007. http://www.guttmacher.org/state center/spibs/spib_RPHS.pdf (accessed July 25, 2007).

Harvard University. "Approval Granted for Harvard Stem Cell Institute Researchers to Attempt Creation of Disease-Specific Embryonic Stem Cell Lines." *Harvard University Gazette Online*, June 8, 2006. http:www.news .harvard.edu/gazette/daily/2006/06/06-stemcell.html (accessed July 21, 2006).

Human Embryo Research Panel, National Institutes of Health. *Report of the Human Embryo Research Panel.* Vol. 1. Bethesda, MD: National Institutes of Health, 1994.

Human Genome Project Information. *What Is Gene Therapy?* Washington, DC: US Department of Energy Office of Science, 2007. http://www.ornl.gov/sci/techresources/Human_Genome/medicine/genetherapy.html (accessed October 7, 2007).

Hyun, Insoo, and Kyu Won Jung. "Human Research Cloning, Embryos, and Embryo-Like Artifacts." *Hastings Center Report* 36, no. 5 (2006): 34–41.

Jefferson, Thomas. *The Life and Selected Writings of Jefferson*. Edited by Adrienne Koch and William Peden. New York: Random House, 1994.

John Paul II. *The Gospel of Life*. New York: Random House, 1995.

Kachigian, Claudia, and Alan R. Felthous. "Court Responses to *Tarasoff* Statutes." *Journal of the American Academy of Psychiatry and the Law* 32 (2004): 263–73.

Kamisar, Yale. "Are Laws against Assisted Suicide Constitutional?" *Hastings Center Report* 23 (1993): 32–41.

———. "Some Non-religious Views against Proposed 'Mercy-Killing Legislation.'" *Minnesota Law Review* 48 (1958): 969–1042.

Kass, Leon R. "Ageless Bodies, Happy Souls: Biotechnology and the Pursuit of Perfection." *New Atlantis*, Spring 2003, pp. 9–28.

———. "'I Will Give No Deadly Drug': Why Doctors Must Not Kill." In *The Case against Assisted Suicide: For the Right to End-of-Life Care*. Edited by Kathleen Foley and Herbert Hendin, 17–40. Baltimore, MD: Johns Hopkins University Press, 2004.

Keely, Lawrence H. *War before Civilization*. New York: Oxford University Press, 1996.

Kevorkian, Jack. *Prescription: Medicide*. Amherst, NY: Prometheus Books, 1991.

Kim, Jong-Hoon, Jonathan M. Auerbach, José A. Rodríguez-Gómez, Iván Velasco, Denise Gavin, Nadya Lumelsky, Sang-Hun Lee, John Nguyen, Rosario Sánchez-Pernaute, Krys Baukiewicz, and Ron McKay. "Dopamine Neurons Derived from Embryonic Stem Cell Function in an Animal Model of Parkinson's Disease." *Nature* 418 (2002): 50–56.

King, Patricia A., and Leslie E. Wolf. "Empowering and Protecting Patients: Lessons for Physician-Assisted Suicide from the African-American Experience." *Minnesota Law Review* 82 (1998): 1015–43.

Kirchoff, Karin T., Prasnath Reddy Anumandla, Kirstine Teresa Foth, Shea Nicole Lues, and Stephanie Ho Gilbertson-White. "Documentation of Withdrawal of Life Support in Adult Patients in the Intensive Care Unit." *American Journal of Critical Care* 13, no. 4 (2004): 328–34.

Kitcher, Philip. "Creating Perfect People." In *A Companion to Genethics*. Edited by Justine Burley and John Harris, 229–42. Oxford: Blackwell, 2002.

Klimanskaya, Irina, Young Chung, Sandy Becker, Shi-Jiang Lu, and Robert

Lanza. "Human Embryonic Stem Cell Lines Developed from Single Blastomeres." *Nature* 444 (2006): 481–85.

Kolata, Gina. "Scientists Bypass Need for Embryo to Get Stem Cells." *New York Times*, November 21, 2007, p. A1.

Korsgaard, Christine M. "Two Distinctions in Goodness." *Philosophical Review* 92 (1983): 169–95.

Krauthammer, Charles. "Stem Cell Vindication." *Washington Post*, November 30, 2007, p. A23.

Kuhse Helga. *The Sanctity-of-Life Doctrine in Medicine: A Critique*. Oxford: Clarendon Press, 1987.

LaFollette, Hugh. "Licensing Parents." *Philosophy and Public Affairs* 9 (1980): 182–97.

The Last of the Mohicans. Directed by Michael Mann. Twentieth Century Fox Film Corporation, 1992.

Ledford, Heidi. "Doubts Raised over Stem-Cell Marker." *Nature* 449 (2007): 647.

Lee, Melinda A., and Linda Ganzini. "Depression in the Elderly: Effect on Patient Attitudes towards Life-Sustaining Therapy." *Journal of the American Geriatrics Society* 40 (1992): 983–88.

Lee, Patrick, and Robert P. George. "The First Fourteen Days of Human Life." *New Atlantis*, Summer 2006, pp. 61–67.

Leshner, Alan I., and James A. Thomson. "Standing in the Way of Stem Cell Research." *Washington Post*, December 3, 2007, p. A17.

Lindsay, Ronald A. "Enhancements and Justice: Problems in Determining the Requirements of Justice in a Genetically Transformed Society." *Kennedy Institute of Ethics Journal* 15 (2005): 3–38.

Lindsay, Ronald A., Tom Beauchamp, and Rebecca Dick. "Hastened Death and the Regulation of the Practice of Medicine." *Washington University Journal of Law and Policy* 22 (2006): 1–28.

———. "The Need to Specify the Difference 'Difference' Makes." *Journal of Law, Medicine and Ethics* 30 (2002): 34–37.

———. "Should We Impose Quotas? Evaluating the 'Disparate Impact' Argument against Legalization of Assisted Suicide." *Journal of Law, Medicine and Ethics* 30 (2002): 6–16.

———. "Slaves, Embryos, and Nonhuman Animals: Moral Status and the Limitations of Common Morality Theory." *Kennedy Institute of Ethics Journal* 15 (2005): 323–46.

———. *Stem Cell Research: An Approach to Bioethics Based on Scientific Naturalism.* Washington, DC: Center for Inquiry, 2006. http://www.centerforinquiry.net/StemCell.pdf

———. "When to Grant Conscientious Objector Status." *American Journal of Bioethics* 7, no. 6 (2007): 25–26.

———. "Why Should We Be Concerned about Disparate Impact?" *American Journal of Bioethics* 6, no. 5 (2006): 23–24.

Loewy, Erich H. "Of Depression, Anecdote, and Prejudice: A Confession." *Journal of the American Geriatrics Society* 40 (1992): 1068–69.

Losey, John E., Linda S. Rayor, and Maureen E. Carter. "Transgenic Pollen Harms Monarch Larvae." *Nature* 399 (1999): 214.

Lydersen, Kari. "Some Muslim Cabbies Refuse Fares Carrying Alcohol." *Washington Post*, October 26, 2006, p. A2.

McHugh, Paul R. "Zygote and 'Clonote'—The Ethical Use of Embryonic Stem Cells." *New England Journal of Medicine* 351 (2004): 209–11.

McKibben, Bill. *Enough: Staying Human in an Engineered Age.* New York: New York Times Books, 2003.

Mehlman, Maxwell J. "The Law of Above Averages: Leveling the New Genetic Enhancement Playing Field." *Iowa Law Review* 85 (2000): 517–93.

———. "Plan Now to Act Later." *Kennedy Institute of Ethics Journal* 15 (2005): 77–82.

———. *Wondergenes.* Bloomington: Indiana University Press, 2003.

Meldrun, Marcia. "'A Calculated Risk': The Salk Polio Vaccine Field Trials of 1954." *British Medical Journal* 317 (1998): 1233–36.

Miller, Henry I., and Gregory Conko. *The Frankenfood Myth: How Protest and Politics Threaten the Biotech Revolution.* Westport, CT: Praeger Publishers, 2004.

"Moralists at the Pharmacy," *New York Times*, April 3, 2005, p. A12.

Mydans, Seth. "After 5 Months of Drama, Brothers' Trials Near End." *New York Times*, December 12, 1993.

National Commission for the Protection of Human Subjects of Biomedical and Behavioral Research. *The Belmont Report: Ethical Principles and Guidelines for the Protection of Human Subject of Research.* Washington, DC: Government Printing Office, 1979.

National Institutes of Health. "Guidelines for Research Using Human Pluripotent Stem Cells." *Federal Register* 65 (2000): 51976–81.

———. *Stem Cell Basics.* 2006. http://stemcells.nih.gov/info/basics (accessed October 23, 2007).

Nestle, Marion. *Safe Food: Bacteria, Biotechnology, and Bioterrorism*. Berkeley, CA: University of California Press, 2003.

New York State Task Force on Life and the Law. *When Death Is Sought: Assisted Suicide and Euthanasia in the Medical Context*. Albany, NY: Health Education Service, 1994.

Obale, Claire C. *Patient Care Decision-Making: A Legal Guide for Providers*. St. Paul, MN: West Group, 1997.

O'Brien, Nancy F. "Pro-Life Official Dismisses New Stem-Cell Announcement as a Sham." *Catholic News Service*, August 28, 2006. http://www.catholicnews.com/data/stories/cns/0604840.htm (accessed October 23, 2007).

Office of the Attorney General. "Dispensing of Controlled Substances to Assist Suicide." *Federal Register* 66 (2001): 56607.

Oregon Department of Human Services. *Eighth Annual Report on Oregon's Death with Dignity Act*, 2006. http://egov.oregon.gov/DHS/ph/pas/docs/year8.pdf (accessed July 4, 2006).

———. *Summary of Oregon's Death with Dignity Act-2006 and Tables 1 and 2*. 2007. http://egov.oregon.gov/DHS/ph/pas/docs/year8.pdf (accessed July 4, 2007).

Orentlicher, David. "The Supreme Court and Assisted Suicide—Rejecting Assisted Suicide but Embracing Euthanasia." *New England Journal of Medicine* 337 (1997): 1236–39.

Parens, Erik. "The Goodness of Fragility: On the Prospect of Genetic Technologies Aimed at the Enhancement of Human Capacities." *Kennedy Institute of Ethics Journal* 5 (1995): 141–53.

———. "Is Better Always Good? The Enhancement Project." In *Enhancing Human Traits: Ethical and Social Implications*, edited by Erik Parens, 1–28. Washington, DC: Georgetown University Press, 1998.

Pellegrino, Edmund D. "Doctors Must Not Kill." *Journal of Clinical Ethics* 3 (1992): 95–102.

———. "The Physician's Conscience, Conscience Clauses, and Religious Belief: A Catholic Perspective." *Fordham Urban Law Journal* 30 (2002): 221–44.

Pence, Gregory E. "Organic or Genetically Modified Food: Which Is Better?" In *The Ethics of Food: A Reader for the 21st Century*, edited by Gregory E. Pence, 116–22. Lanham, MD: Rowman and Littlefield, 2002.

Phillips, Don. "Federal Speed Limit, Set in 1974, Repealed." *Washington Post*, November 29, 1995, pp. A1, A16.

Photonics.com. "Laser Inventor Dies at 79." 2007. http://www.photonics.com/content/news/2007/May/9/87661.aspx (accessed September 29, 2007).

Plotz, David. "Wake Up, Little Susie: Can We Sleep Less?" *Slate*, March 7, 2003. http://www.slate.com/id/2079113/ (accessed October 7, 2007).

Posner, Richard A. *Aging and Old Age*. Chicago: University of Chicago Press, 1995.

Potrykus, Ingo. *Comments on the World Development Report 2008*. 2007. http://www.goldenrice.org/Conent4-Info/info7_actuality.htm (accessed September 30, 2007).

President's Commission for the Study of Ethical Problems in Medicine and Biomedical Behavioral Research. *Deciding to Forego Life-Sustaining Treatment*. Washington, DC: Government Printing Office, 1983.

———. *Defining Death: Medical, Legal and Ethical Issues in the Determination of Death*. Washington, DC: Government Printing Office, 1981.

President's Council on Bioethics. *Alternative Sources of Human Pluripotent Stem Cells*. Washington, DC: President's Council on Bioethics, 2005.

———. *Human Cloning and Human Dignity*. Washington, DC: President's Council on Bioethics, 2002.

———. *Monitoring Stem Cell Research*. Washington, DC: President's Council on Bioethics, 2004. http://www.bioethics.gov/reports (accessed July 17, 2006 and October 23, 2007).

Preston, Richard. "An Error in the Code." *New Yorker*, August 23, 2007.

Quill, Timothy E., Rebecca Dresser, and Dan W. Brock. "The Rule of Double Effect—A Critique of Its Role in End-of-Life Decision Making." *New England Journal of Medicine* 337 (1997): 1768–71.

Rachels, James. *The End of Life*. New York: Oxford University Press, 1986.

Raffensperger, Carolyn, and Joel Tickner, eds. *Protecting Public Health and the Environment: Implementing the Precautionary Principle*. Washington, DC: Island Press, 1999.

Rawls, John. *A Theory of Justice*. Cambridge, MA: Harvard University Press, 1971.

Raz, Joseph. *The Morality of Freedom*. New York: Oxford University Press, 1986.

Regan, Tom. *The Case for Animal Rights*. Berkeley: University of California Press, 1983.

Reinan, John. "Taxi Proposal Gets Sharp Response." *Minneapolis–St. Paul Star Tribune*, February 27, 2007, p. 1.

Rennison, Callie Marie, and Sarah Welchans. *Intimate Partner Violence.* Washington, DC: US Department of Justice, 2000. http://www.ojp.usdoj.gov/bjs/pub/pdf/ipv.pdf (accessed July 14, 2007).

Samuelson, William, and Richard Zeckhauser. "Status Quo Bias in Decision Making." *Journal of Risk and Uncertainty* 1 (1988): 7–59.

Sandel, Michael J. *The Case against Perfection: Ethics in the Age of Genetic Engineering.* Cambridge, MA: Harvard University Press, 2007.

Sankula, Sujatha. *Quantification of the Impacts on U.S. Agriculture of Biotechnology-Derived Crops Planted in 2005.* Washington, DC: National Center for Food and Agricultural Policy. 2006. http://ncfap.org/whatwedo/pdf/2005biotechExceSummary.pdf (accessed September 30, 2007).

Schlissel, Lillian. *Conscience in America: A Documentary History of Conscientious Objection in America, 1757–1967.* New York: E. P. Dutton, 1968.

Scott, Susan E., and Mike J. Wilkinson. "Low Probability of Chloroplast Movement from Oilseed Rape (*Brassica napus*) into Wild *Brassica rapa.*" *Nature Biotechnology* 17 (1999): 390–93.

Sears, Mark, Richard L. Hellmich, Diane E. Stanley-Horn, Karen S. Oberhauser, John M. Pleasants, Heather R. Mattila, Blair D. Siegfried, and Galen P. Dively. "Impact of Bt Corn Pollen on Monarch Butterfly Populations: A Risk Assessment." *Proceedings of the National Academy of Sciences* 98 (2001): 11937–42.

Seligman, Martin E. P. *Learned Depression.* New York: Random House, 1991.

Silver, Lee M. *Challenging Nature: The Clash between Biotechnology and Spirituality.* New York: Harper Perennial, 2007.

Singer, Peter. *Animal Liberation.* New York: New York Review Books, 1975.

———. *In Defense of Animals: The Second Wave.* Oxford: Blackwell, 2006.

Snow, Tony. Press Briefing, the White House, July 18, 2006. http://www.whitehouse.gov/news/releases/2006/07/20060718.html (accessed October 23, 2007).

———. Press Briefing, the White House, July 24, 2006. http://www.whitehouse.gov/news/releases/2006/07/20060724-4.html (accessed October 23, 2007).

Southern Baptist Convention. *On Human Species-Altering Technologies. Resolution no.* 7. Greensboro, NC: Southern Baptist Convention, 2006. http://www.sbcannualmeeting.net/sbc06/resolutions/sbcresolution-06.asp?ID=7 (accessed October 18, 2007).

Spencer, Rowena. "Parasitic Conjoined Twins: External, Internal (Fetuses in Fetu and Teratomas), and Detached (Arcardiacs)." *Clinical Anatomy* 14 (2001): 428–44.

Spindelman, Marc S. "Legislating Privilege." *Journal of Law, Medicine and Ethics* 30 (2001): 24–33.

Stanley, Barbara, Michael Stanley, Jeannine Guido, and Lynn Garvin. "The Functional Competency of the Elderly at Risk." *Gerontologist* 28 (3 suppl.) (1988): 53–58.

Stein, Alexander J., H. P. S. Sachdev, and Martin Quinn. "Potential Impact and Cost-Effectiveness of Golden Rice." *Nature Biotechnology* 24 (2006): 1200–1201.

Stein, Rob. "Pharmacists' Rights at Front of New Debate." *Washington Post*, March 28, 2005, p. A1.

Steinbock, Bonnie. "The Morality of Killing Human Embryos." *Journal of Law, Medicine and Ethics* 34 (2006): 26–34.

Sullivan, Mark D., and Stuart J. Younger. "Depression, Competence and the Right to Refuse Lifesaving Medical Treatment." *American Journal of Psychiatry* 151 (1994): 971–78.

Swartz, Martha S. "'Conscience Clauses' or 'Unconscionable Clauses': Personal Beliefs versus Professional Responsibilities." *Yale Journal of Health Policy, Law, and Ethics* 6 (2006): 269–350.

Thomson, James A., Joseph Itskovitz-Eldor, Sander S. Shapiro, Michelle A. Waknitz, Jennifer J. Swiergiel, Vivienne S. Marshall, and Jeffrey M. Jones. "Embryonic Stem Cell Lines Derived from Human Blastocysts." *Science* 282 (1998): 1145–47.

Tolle, Susan W., Virginia P. Tilden, Linda L. Drach, Erik K. Fromme, Nancy A. Perrin, and Katrina Hedberg. "Characteristics and Proportion of Dying Oregonians Who Personally Consider Physician-Assisted Suicide." *Journal of Clinical Ethics* 15 (2004): 111–18.

Tomalin, Claire. *Thomas Hardy*. New York: Penguin Press, 2007.

Tsien, Joe Z. "Building a Brainier Mouse." *Scientific American*, April 2000, pp. 62–68.

Turnbull, Colin. *The Mountain People*. New York: Simon & Schuster, 1972.

Turner, Danielle C., Trevor W. Robbins, Luke Clark, Adam R. Aron, Jonathan Dowson, and Barbara J. Sahakian. "Cognitive Enhancing Effects of Modafinil in Healthy Volunteers." *Psychopharmacology* 165 (2003): 260–69.

Union of Concerned Scientists. *Food and Environment: A Special Note to Organic*

Consumers, 2007. http://www.ucsusa.org/food_and_environment/genetic_engineering/note-to-organic-consumers.html (accessed August 18, 2007).

US Congress. House. Committee on the Judiciary. *Assisted Suicide in the United States*. 104th Cong., 2nd sess., 1996.

US Department of Agriculture. Organic Foods Production Act Provisions. *Code of Federal Regulations*, title 7, secs. 205.1–205.690 (2007).

US Department of Health and Human Services (HHS). *A Nation's Shame: Fatal Child Abuse and Neglect in the United States*. Washington, DC: Government Printing Office, 1995.

———. *The Third National Incidence Study of Child Abuse and Neglect*. Washington, DC: Government Printing Office, 1996.

Vaux, Kenneth. "The Heart Transplant: Ethical Dimensions." *Christian Century* 85 (1968): 353–56.

Veatch, Robert M. *Transplantation Ethics*. Washington, DC: Georgetown University Press, 2000.

Wade, Nicholas. "Clinics Hold More Embryos Than Had Been Thought." *New York Times*, May 9, 2003, p. A24.

Weiss, Rick. "Fungus Infected Woman Who Died after Gene Therapy." *Washington Post*, August 17, 2007, p. A10.

———. "Mature Human Embryos Created from Adult Skin Cells." *Washington Post*, January 18, 2008, p. A1.

———. "Probe into Tainted Rice Ends." *Washington Post*, October 6, 2007, p. A2.

———. "Scientists See Potential in Amniotic Stem Cells." *Washington Post*, January 8, 2007, p. A1.

Weissman, Irving L. "Stem Cells—Scientific, Medical and Political Issues." *New England Journal of Medicine* 346 (2002): 1576–79.

White, Mathew. "Conscience Clauses for Pharmacists: The Struggle to Balance Conscience Rights with the Rights of Patients and Institutions." *Wisconsin Law Review* (2005): 1611–48.

Willer, Helga, and Minou Yussefi, eds. *The World of Organic Agriculture: Statistics and Changing Trends*, 9th ed. Bonn, Germany: International Federation of Organic Agriculture Movements, 2007. http://orgprints.org/10506/01/willer-yussefi-2007-p1-44.pdf (accessed August 20, 2007).

Wolf, Susan M. "Physician-Assisted Suicide, Abortion and Treatment Refusal." In *Physician-Assisted Suicide*, edited by Robert F. Weir, 167–201. Bloomington: Indiana University Press, 1997.

Wu, Guoyao, Fuller W. Bazer, Timothy A. Cudd, Cynthia J. Meininger, and Thomas E. Spencer. "Maternal Nutrition and Fetal Development." *Journal of Nutrition* 134 (2004): 2169–72.

Yoon, Carol Kaesuk. "Altered Corn May Imperil Butterfly, Researchers Say." *New York Times*, May 20, 1999, p. A1.

Yu, Junying, Maxim A. Vodyanik, Kim Smuga-Otto, Jessica Antosiewicz-Bourget, Jennifer L. Frane, Shulan Tian, Jeff Nie, Gudrun A. Jonsdottir, Victor Ruotti, Ron Stewart, Igor I. Slukin, and James A. Thomson. "Induced Pluripotent Stem Cell Lines Derived from Human Somatic Cells." *Science* 318 (2007): 1917–20.

LEGAL CASES

Albemarle Paper Co. v. Moody, 422 U.S. 405 (1975).

Bowen v. Roy, 476 U.S. 693 (1986).

Bradfield v. Roberts, 175 U.S. 291 (1899).

Bruff v. North Miss. Health Services, Inc., 244 F.3d 495 (5th Cir. 2001).

Clackamas Gastroenterology Associates, P.C. v. Wells, 538 U.S. 440 (2003).

Compassion in Dying v. Washington, 79 F.3d 790 (9th Cir. 1995) *rev'd and remanded sub nom. Washington v. Glucksberg*, 521 U.S. 702 (1997).

Eisenstadt v. Baird, 405 U.S. 438 (1972).

Endres v. Indiana State Police, 349 F.3d 922 (7th Cir. 2003).

Gonzales v. Oregon, 546 U.S. 243 (2006).

Griggs v. Duke Power, 401 U.S. 424 (1971).

Griswold v. Connecticut, 381 U.S. 479 (1965).

In re Quinlan, 70 N.J. 10 (1976).

Loving v. Virginia, 388 U.S. 1 (1967).

People v. Kevorkian, 210 Mich.App. 601, 534 N.W. 2d. 172 (1995).

Planned Parenthood v. Casey, 505 U.S. 833 (1992).

Shelton v. University of Med. and Dentistry of New Jersey, 223 F.3d 220 (3rd Cir. 2000).

Tarasoff v. Regents of the University of California, 551 P.2d 334 (Cal. 1976).

Trans World Airlines v. Hardison, 432 U.S. 63 (1977).

Vacco v. Quill, 521 U.S. 793 (1997).

Washington v. Glucksberg, 521 U.S. 702 (1997).

Weber v. Roadway Express, 199 F.3d 270 (5th Cir. 2000).
Welsh v. United States, 398 U.S. 333 (1970).

STATUTES

Federal

Balanced Budget Downpayment Act (Dickey Amendment), U.S. Statutes at Large 110 (1996): 26.
Civil Rights Act of 1964, *U.S. Code*, vol. 42, secs. 2000e–2000e-15 (2003).
Civil Rights Act of 1991, U.S. Code, vol. 42, sec. 2000e-2(k)(l)(A)(i)(2003).
Health Programs Extension Act of 1973 (Church Amendment), U.S. Code, vol. 42, sec. 300a-7 (2003).
Patient Self-Determination Act of 1990, U.S. Code, vol. 42, sec. 1395cc(f)(2003).

State

Health Care Right of Conscience Act, Illinois Compiled Statutes (2006) secs. 745–70.
Health Care Rights of Conscience Act of 2004, Mississippi Code (2007) sec. 41-107-7.
Oregon Death with Dignity Act, Oregon Revised Statutes (2003) secs.127.800–995.

Index